EXPERIMENTAL PHYSICAL
CHEMISTRY

EXPERIMENTAL PHYSICAL CHEMISTRY

BY

W. G. PALMER, M.A., Sc.D., D.Sc.

Fellow of St John's College, Cambridge
formerly University Lecturer in Chemistry
in the University of Cambridge

SECOND EDITION

CAMBRIDGE
AT THE UNIVERSITY PRESS
1962

CAMBRIDGE UNIVERSITY PRESS
Cambridge, New York, Melbourne, Madrid, Cape Town, Singapore, São Paulo, Delhi

Cambridge University Press
The Edinburgh Building, Cambridge CB2 8RU, UK

Published in the United States of America by Cambridge University Press, New York

www.cambridge.org
Information on this title: www.cambridge.org/9780521104951

First edition 1941
Reprinted with corrections 1946, 1949, 1954
Second edition 1962
This digitally printed version 2009

A catalogue record for this publication is available from the British Library

ISBN 978-0-521-05903-9 hardback
ISBN 978-0-521-10495-1 paperback

Contents

Preface

In planning the course of work contained in this book I have tried to keep the following considerations before me. First, that students both at Universities and in schools cannot often afford to devote more than limited periods to practical work in any one scientific subject. The provision of detailed instructions is therefore imperative, in order that a satisfying result may be reached without a discouraging waste of time. Secondly, that while being capable of yielding reasonable accuracy, the apparatus ought to be simple enough to be assembled or constructed by the students themselves from ordinary laboratory equipment. Further, that an approach to many important principles through the laboratory, rather than by way of the lecture, more easily stimulates interest, and more certainly ensures comprehension.

Most of the exercises have for many years past formed part of a course of physical chemistry at the University Chemical Laboratory at Cambridge. While much of the course has been found suitable for beginners in the subject, who are candidates for an Honours Degree through the Natural Sciences Tripos, a number of experiments are described which are appropriate to more advanced students. The limitations of time (which for beginners usually consists of periods not longer than three hours) have influenced the choice and nature of the experiments, necessitating, even if it were not desirable upon other grounds, the simplicity of home-made apparatus, and sometimes the substitution of rough-and-ready means of observation for less direct methods, which would overtax both the time and the experience of the beginner. I hope that a course of this type may interest teachers and students in other

Universities and Colleges, and that the simplicity of the equipment required in many of the experiments may render
them suitable for Higher Certificate and Scholarship forms in
schools.

While the notes introducing each chapter are not intended
to replace the use of adequate text-books on the theory of
the subject, it may be hoped that if the student can readily
refer to principles while at work on a problem, he will more
easily interweave practical wisdom with theoretical knowledge, and come to realize that the former is the real founda-
tion of the latter. The inclusion of completely worked examples
of nearly all the experiments has been made a regular feature.
These examples are all based upon data obtained in the
Cambridge laboratory from actual materials and apparatus
described in the text. They will justify their inclusion if they
elucidate difficulties in the most direct way, demonstrate the
possible accuracy attainable, and assist in encouraging an
orderly and significant exposition of the data. I should be
grateful to receive helpful suggestions from teachers and
students who may use the book, but I venture to counter the
criticism that the practical instructions are so detailed as to
impugn the students' initiative, by remarking that if a beginner spends the greater part of his limited laboratory period
in merely prospecting an experiment, with no satisfying result,
discouragement will commonly overshadow all that he may
well have learnt in manipulation.

In a laboratory where all one's fellow-teachers are also
one's friends and well-wishers, it is very difficult to offer
thanks to some rather than to all, but I must specially
mention the help I have received in discussion on Chapter VIII
(Surface Chemistry) from Professor Lennard-Jones, and from
Mr G. E. Briggs (of the Cambridge Botany School); from Dr
R. C. Evans on Chapter II (The Properties of Crystals), and

from Mr A. J. Berry and Dr T. P. Hoar on Chapters v, vi, and vii (Thermochemistry, Ionization and Reaction Velocity). All of these friends have been good enough to read parts of the book in MS. or proof. I am indebted to my son, R. G. Palmer, for the estimations upon which fig. 12 is based. I also feel bound to pay homage to the memory of Dr H. J. H. Fenton, whose teaching, with its masterly blend of enthusiasm, scholarship, and scepticism, it was a rare privilege to receive. To the staff of the Cambridge University Press I also owe a large debt of gratitude, both for their patient solicitude, and their unrivalled expertness, which has so greatly improved the tenor of the book.

W. G. P.

June 1941

Largely through the agency of those who, having found *Experimental Physical Chemistry* acceptable in their teaching, have afforded me the assistance of their friendly and constructive criticisms, I have been enabled to introduce some minor emendations and improvements into this new impression, notably on pp. 32, 238 and 245.

W. G. P.

May 1951

Preface to Second Edition

The principal task in preparing a new edition has been the reviewing of data and references in the light of recent progress in chemistry. In consequence many of the data have needed some adjustment, and new references have superseded a majority of those formerly appearing. Where changes have been made in the text it is hoped that improvement has resulted.

W. G. P.

Nov. 1961

SYMBOLS AND CONVENTIONS

List of symbols

A area, molecular surface area, gas-density correction.

a activity (especially of electrolytes).

α degree of (electrolytic) dissociation.

β conductance ratio.

C, c molecular concentration; number of components (phase rule); equivalent concentration.

c specific heat, molecular heat.

D, d density (all states of matter).

E e.m.f.; electrode potential; energy (general).

η coefficient of viscosity.

F force; number of degrees of freedom (phase rule).

F faraday.

f activity coefficient.

G free energy ($H - TS$); temperature coefficient of reaction velocity.

g osmotic coefficient (electrolytes).

H heat content (ΔH, heat change at constant pressure).

K, K_c, K_p equilibrium constant (concn. or press. terms).

κ specific conductance.

L, l latent heat; latent heat per g.

l ionic mobility (ionic conductance); liquid phase (phase rule).

Λ equivalent conductance.

M molecular weight.

μ ionic strength.

N, n number of (g.) molecules

N_A molar fraction (of A).

P, p pressure; number of phases (phase rule); partial pressure.

ϕ coefficient of fluidity.

R gas constant per mol. per degree.

R, ρ resistance; specific resistance.

S entropy; percentage of solute in solution.

s solid phase (phase rule).

σ molecular cross-section; surface tension (surface free energy).

T, t absolute temperature; temperature Centigrade; time.

τ time interval.

U total energy (ΔU, heat change at constant volume).

V, v volume, dilution; partial volume.

V., mV. volt; millivolt.

W work.

w weight.

z ionic charge.

Z collision frequency.

Mathematical and chemical signs

$\simeq$ approximately equal to.

$\Delta (A)$ change of (A).

$\log x = \log_{10} x$.

$\quad \log_{10} e = 0.4343$.

$\ln x = \log_e x$.

$\quad \log_e 10 = 2.303$.

$[A]$ concentration of (A).

xN x normal solution.

b.p. boiling-point.

f.p. freezing-point.

m.p. melting-point.

Fundamental constants (chemical scale)*

Avogadro number, N, 6.023×10^{23}.

Faraday, F, 96,494 coulombs (abs.).

Electronic charge, e, 4.803×10^{-10} e.s.u.

Gas constant, R, 0.08205 litre-atmosphere per mol. per degree; 1.986 calories per mol. per degree; 8.314 joules per mol. per degree.

Molecular volume of a perfect gas at N.T.P. 22.415 litres.

Absolute zero of temperature $-273.16°$.

* Cohen *et al.*, *Rev. Mod. Phys.*, **27**, 363 (1955).

REFERENCES TO LITERATURE

Acta. chem. Scand. Acta chemica Scandinavica.
Amer. Min. American Mineralogist.
Ann. Physik. Annalen der Physik (Leipzig).
Arch. phys. biol. Archives de physique biologique.
Arch. Sci. phys. nat. Archives des Sciences physiques et naturelles (Geneva).
Atti. R. Accad. Lincei. Atti (Rendiconti) della reale Accademia dei Lincei.
Ber. Berichte der deutschen chemischen Gesellschaft.
Biochem. Z. Biochemische Zeitschrift.
Bull. Soc. chim. France. Bulletin de la Société chimique de France.
Can. J. Chem. Canadian Journal of Chemistry.
Chem. Soc. Abstr. Abstracts of the Chemical Society or British Chemical Abstracts.
Gazzetta. Gazzetta chimica italiana.
I.C.T. International Critical Tables.
Ind. Eng. Chem. Industrial and Engineering Chemistry.
Inst. Metals. Ann. Diag. Institute of Metals, Annotated Diagrams.
J. Amer. Chem. Soc. Journal of the American Chemical Society.
J. Chem. Physics. Journal of Chemical Physics.
J. Chem. Soc. Journal of the Chemical Society (London).
J. chim. physique. Journal de chimie physique (Geneva).
J. Gen. Physiol. The Journal of General Physiology (Baltimore).
J. Physical Chem. The Journal of Physical Chemistry.
J. prakt. Chem. Journal für praktische Chemie.
J. Soc. Chem. Ind. Journal of the Society of Chemical Industry.
Kolloid-Z. Kolloid-Zeitschrift.
Monograph Amer. Chem. Soc. Monographs of the American Chemical Society (Reinhold Publishing Corp. U.S.A.).
Natl. Bur. Stand. National Bureau of Standards (U.S.A.).
Natl. Bur. Stand. J. Res. National Bureau of Standards (U.S.A.), Journal of Research.
Phil. Mag. Philosophical Magazine (London, Edinburgh and Dublin).
Phil. Trans. Philosophical Transactions of the Royal Society.
Proc. phys. Soc. Proceedings of the Physical Society.
Proc. Roy. Soc. Proceedings of the Royal Society (London).
Quart. Rev. Quarterly Reviews (Chemical Society).
Rec. Trav. chim. Recueil des Travaux chimiques des Pays-Bas.
Rev. Mod. Phys. Reviews of Modern Physics.
Trans. Electrochem. Soc. Transactions of the Electrochemical Society.
Trans. Faraday Soc. Transactions of the Faraday Society.
Z. anal. Chem. Zeitschrift für analytische Chemie.
Z. angew. Chem. Zeitschrift für angewandte Chemie.
Z. anorg. Chem. Zeitschrift für anorganische und allgemeine Chemie.
Z. Elektrochem. Zeitschrift für Elektrochemie.
Z. Krist. Zeitschrift für Kristallographie.
Z. physikal. Chem. Zeitschrift für physikalische Chemie, Stöchiometrie und Verwandtschaftslehre.

Introduction

NOTES ON WEIGHING, CALIBRATION, AND THE USE OF GRAPHICAL METHODS

(1) Weighing

To weigh 0·5 g. of precipitate to 1 mg., or 1 in 500, when it is contained in a crucible weighing 20 g., demands an accuracy of 1 in 20,000 in the balance, which should therefore be treated with the care and attention due to delicate apparatus. The sensitiveness of a balance, measured by the displacement of the beam by 1 mg. difference in load, depends principally on the inertia, i.e. size, of the beam, and, less markedly, on the total load. A micro-balance, used for weighing minute masses, is therefore literally a small balance.

If a balance of the enclosed type is of good quality, kept in a suitable room at even temperature and not exposed to direct sunlight, it may be assumed that mechanical defects, such as unequal length of beam, are absent. It may, however, be remarked that weighing by difference, which is almost always the method of chemistry, does not of itself eliminate the effect of such a fault, unless the nature of the determination in hand implies that the weighings will be used in ratio; thus the error is eliminated in a cryoscopic determination of molecular weight when the solvent and the solute are both weighed on the same pan of the same balance, but not when the solvent is measured by volume. In the latter case, however, the error will usually be merged in the probably greater errors in volume measurement. It is hardly worth while to calibrate good-quality modern weights; if this is attempted, it is probably better regarded as a test of the operator's skill in exact weighing than as a test of the accuracy of the weights.

It is advisable to acquire the habit of regarding the scale

over which the pointer swings as graduated thus (or in the opposite sense):

0	5	10	15	20

and not to place the zero graduation in the centre. On the plan advocated the resting-point is found simply by dividing the sum of the readings of the extremities of the swing by two, and confusing differences of sign will not arise. Thus if the swings give 6·5, 15·1 and 7·0, the resting-point is

$$\tfrac{1}{2}(6\cdot75 + 15\cdot1) = 10\cdot9.$$

With a little practice this calculation can be correctly made without the aid of pencil and paper.

Before proceeding to an accurate weighing, the resting-point of the unloaded balance should always be checked; only if this point does not lie within the limits 7 to 13 does the balance need adjustment, by appropriate movement of one of the small screws at the ends of the beams. Much time is saved in the final adjustment of the rider if the sensitiveness (defined above) is approximately known. To determine this, set the rider at its zero position (where it should always be found at the beginning of a weighing, and when the balance is out of use), and then at the 1 or 2 mg. mark, observe the swings and calculate the resting-point for each position; the difference gives the required sensitiveness. It may be assumed that for moderate loads the sensitiveness is constant, but the variation can, of course, be tested if desired by carrying out the above operation with different equal loads on the pans.

Before starting an exact weighing, demanding considerable time, it is worth while to consider whether the accuracy attainable in other parts of the work merits such exactitude. A 'rough', i.e. open, balance weighing to 1 cg. should not be despised, and its employment in suitable instances may avoid much waste of time. Such a balance should always be used as a preliminary when it is desired, as in a freezing-point determination, to place in the weighing tube a pre-determined amount of substance, and for loads greater than 100 g.

It is perhaps unnecessary to stress the safeguard of double-checking the weights used, by adding on the pan, and subtracting in the box. Some boxes of weights are provided with only three compartments for the four pieces 500, 200 and two 100 mg. In this case mistakes are avoided by keeping the similarly shaped 100 mg. pieces separate, and the unlike 500 and 200 mg. in the same section. If necessary adopt the same arrangement for the 50, 20 and 10 mg. pieces.

(2) Calibration and standardization

Apparatus which may be expected to need calibration for use in work of even moderate accuracy comprises (a) volume-measuring apparatus, (b) temperature-measuring instruments.

(a) *Volume-measuring apparatus*

A distinction must be drawn between delivery apparatus, such as pipettes and burettes for liquids, and apparatus designed to contain a known volume. Calibration of pipettes is effected by weighing a charge of liquid delivered into a stoppered weighing bottle, the exact density of the liquid at the temperature concerned being ascertained from tables. Wherever possible the calibration should be made with the liquid to be subsequently used in the pipette, as differences of surface properties and viscosity lead to different drainage and shape of meniscus; mercury should never be used in standardizing pipettes for other liquids. A calibration is only strictly valid for the temperature at which it has been made, but no error appreciable in ordinary work is introduced by neglect of this point, if only small temperature variations are allowed. Volume-containing apparatus, such as graduated flasks, are calibrated by filling with liquid and weighing, on a suitable balance. Here drainage does not come in question, and difference in shape of meniscus is usually a very small fraction of the whole volume, so that water may always be used.

Calibration of burettes for liquid delivery. Burettes for delivery of liquid may be standardized (for aqueous liquids) by the following method. An auxiliary pipette of the form shown in fig. 1 is prepared. The bulb should have a capacity somewhat less than the volume steps in which it is desired to calibrate the burette; steps of 5 ml. are not too large for a burette of good quality. The delivery tube of the pipette is closed by inserting a small glass bead B in the rubber connexion c (a spring-clip should *not* be used). The pipette is firmly attached to the delivery tip of the burette by a stout rubber union (in which it is desirable that the ends of the glass should touch), and held upright by the wire w.

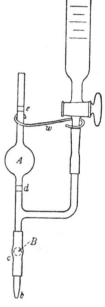

Fig. 1

Water is flushed from the burette through the connexions to remove all air-bubbles. The bulb A is emptied by squeezing the rubber round the bead B and thus opening a small channel; no air-bubble must remain in either connexion. An arbitrary level is marked below the bulb A at d by surrounding the tube with a narrow band of gummed paper. The burette is now filled, the water level brought as exactly as possible to the zero mark, and the burette tap then shut. Water is then run from the pipette (by opening the bead closure) exactly to the mark d, time being allowed for the full drainage of the bulb. The pipette is now refilled from the burette by allowing the water to descend slowly to the 5 ml. mark; the setting on the burette (which need not be exactly at 5 ml.) is read, and the upper level of the water in the pipette marked on the stem at e by the use of gummed paper as before. The contents of the pipette are now discharged to the mark d into a previously weighed stoppered bottle. Another portion of water is now allowed to flow from the burette exactly to the upper mark e, the burette setting noted, and the pipette then discharged as before into the bottle.

Proceed in this way until the whole content of the burette has been run out in equal portions. The total weight of the water collected in the bottle being ascertained by a second weighing, its total volume is found by taking its temperature and using the density obtained from tables. The capacity between the marks on the pipette is now exactly known, and hence the true volume corresponding to the burette readings. A graph is made on squared paper relating readings to true volume, and the points joined by straight lines. Corrections for intermediate readings can then be read from the graph.

Example

Total weight of ten pipette charges = 50·440 g.

Temperature = 18·0°.

$$\text{Pipette capacity} = \frac{5·044}{0·99878} = \frac{5·044}{1-0·00122}$$

$$= 5·044\,(1+0·00122)*$$

$$= 5·049 \text{ ml.}$$

Burette reading	True volume ($n \times 5·049$)	Correction
5·00	5·05	+0·05
10·05	10·10	+0·05
15·10	15·15	+0·05
20·20	20·20	0·00
25·21	25·25	+0·04
etc.	etc.	etc.

Density of water (g./ml.)		Density of water (g./ml.)	
°C.		°C.	
15	0·99929	19	0·99859
16	0·99913	20	0·99839
17	0·99897	21	0·99818
18	0·99878	22	0·99796

Calibration of gas-holding apparatus. The capacity of glass apparatus intended to contain gas should be calibrated with

* When x is small, $(1-x)^{-1} = 1+x$.

mercury, when the connexions allow, and the size is small, but more usually it will be necessary to calibrate by ascertaining the pressure change when *dry* air, at atmospheric pressure, is expanded from a standard capacity into the highly exhausted apparatus. For maximum accuracy the standard volume should be nearly equal to the volume to be standardized. Gas burettes may be calibrated by a method analogous to that given above for burettes delivering liquid. A small glass pipette, of 2–5 ml. capacity, with capillary tubes (fig. 2), is first calibrated with mercury between marks on the tubes. The pipette is then connected with the gas burette as shown, and filled with mercury to the upper mark with the aid of the levelling funnel *F*. The gas burette is filled with dry air through the side-tube *t* and the mercury set to its lowest graduation and levelled. The side-tube being now closed, air is drawn from the burette into the pipette to the lower mark by lowering the funnel *F*. After again levelling the mercury in both pipette and burette, the reading on the latter is noted. The air is now expelled from the pipette through the opened side-tube and the above operation repeated until the whole of the air has been taken from the burette.

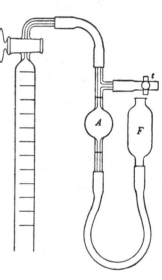

Fig. 2

(b) *The calibration of thermometers, and thermo-couples*

As fixed points the invariant systems provided by solid-liquid mixtures, under atmospheric pressure, are always to be preferred to boiling-points, on account of the insensitiveness of the former to pressure change, and also because in general crystalline solids are far more easily purified than liquids.

Transition points of hydrated salts are eminently suitable as fixed points in the range near room temperature, as the required systems are very readily set up, and the temperature, once established, remains constant, even in simple apparatus, for an exceptionally long period, usually over 1 hr.:

$0°$. Ice–water. For procedure, see exp. c (i), p. 104.

$32\cdot383°$. $Na_2SO_4 \cdot 10H_2O \rightleftharpoons Na_2SO_4 + 10H_2O$.* Pure sodium sulphate (e.g. *Analar* brand) must be used. The method of exp. b (i), p. 73, should be followed.

$48\cdot45°$. $Na_2S_2O_3 \cdot 5H_2O \rightleftharpoons Na_2S_2O_3 \cdot 2H_2O + 3H_2O$.†

$50\cdot674 \pm 0\cdot005°$. $NaBr \cdot 2H_2O \rightleftharpoons NaBr + 2H_2O$.‡

The solubility diagram for sodium bromide is very simple,§ while that for the thiosulphate is exceptionally complex†. On this account the sodium bromide system would be preferred, were it not for the extreme difficulty of purifying the salt from chloride (Richards and Wells).

To use the thiosulphate system, proceed as follows. 50 g. of the pure pentahydrate (*Analar* brand) are placed in a boiling tube and kept at about $60°$ until an easily stirred mixture of solid and solution is produced. The tube is closed with a grooved cork carrying the thermometer, and is transferred to a larger tube serving as an air-jacket (see fig. 34), which may, for greater accuracy, be immersed in a water-bath regulated to $40°$. The mixture is stirred regularly as the temperature falls to about $47°$, when it is seeded with a small quantity of the pentahydrate. After inoculation the stirring is continued vigorously, when the temperature rises rather slowly to a maximum at $48\cdot45°$, where it remains constant for

* Redlich and Loeffler, *Z. Elektrochem.* **36**, 716 (1930).
† Picon, *Bull. Soc. chim. France*, **35**, 1097 (1924).
‡ Richards and Wells, *Z. physikal. Chem.* **56**, 348 (1906).
§ Abegg, *Handbuch*, **2**, Abt. I (Group I), p. 236; also Gmelin, *Handbuch*, Aufl. 8, Syst.-Nr. 21, Sodium, p. 420.

over an hour, if the mixture is stirred occasionally to prevent it from clotting into a hard mass:

80·22°. The melting-point of naphthalene*. For method see p. 96.

100°. Use a steam-distillation method, as on p. 92.

For temperatures over 100° the following melting-points of metals are to be recommended; commercial pure tin and zinc are suitable, but a specially pure quality is required for lead:

183 ± 0·3°. The tin-lead eutectic.† The required mixture contains 61·9 % by weight of tin.

231·9°. The melting-point of tin.‡

327·4°. The melting-point of lead.‡

419·5°. The melting-point of zinc.‡

For ordinary work the metals may be melted in deep crucibles under a layer of charcoal, and the temperature arrest on cooling registered; very little under-cooling is usually observed (see exp. (*a*), p. 97). The eutectic mixture should be occasionally stirred (with a thin carbon-arc pencil) during cooling. For very accurate standardization the metals must be manipulated in an inert atmosphere.

(3) The use of graphical methods

The portrayal of a series of experimental results in a graph is too well known a method to need description here, but the graph is too often drawn outside the laboratory after the experiment is completed, and possibility of modification or confirmation has passed. When an experiment involves the observation of a continuous change (e.g. cooling or heating curves, freezing-points in a molecular-weight determination, titration of samples in a reaction-velocity experiment), each successive reading should be plotted on squared paper directly it is taken.

* Timmermans, *Physico-chemical Constants of Pure Organic Compounds*, Elsevier, 1950, p. 178.

† Raynor, *Inst. Metals Ann. Diag.* No. 6, 1954.

‡ *Natl. Bur. Stand., Circular C* 500 (1952).

It is only from the evidence of a contemporaneous graphical record that the course of an experiment involving continuous variation can be intelligently followed, expected changes recognized, and, no less important, the source of novel effects explored in time to be either confirmed or rejected as spurious. When the magnitude of the probable error in the readings is known, this should be expressed in the graph by using lines, ellipses or circles instead of points.

A fundamental simplicity in natural law may legitimately encourage an expectation of discovering numerous linear relationships, but it is well to make sure in a particular case that the linear appearance of a graph is not due merely to an inappropriate choice of scale, or, for example, to the fact that $\log x$ varies more slowly than x.

In the author's opinion firm freehand drawing is superior to any form of flexible 'spline' in the production of justly drawn curvilinear graphs, but a length of 'composition' gas piping, $\frac{3}{16}$ or $\frac{1}{4}$ in. external diameter, which can be very easily bent to follow even high curvatures, forms a valuable guide if the graph is to be inked over. The best 'lie' of a suspected linear graph cannot be properly assessed by using an opaque ruler, and only indifferently by a transparent one. A better method consists in sliding the paper on which the points are marked under a length of thread stretched tightly between two drawing pins, until a satisfactory setting is reached; the correct ordinate and abscissa are noted and the line subsequently drawn through these points. A similar device can be recommended for 'tangenting' a curve (exp., p. 248).

Chapter I

THE DENSITIES OF GASES AND VAPOURS

The absolute density of a gas is defined as the mass in grams of 1 l. at N.T.P., and the relative density as the ratio of the density to that of hydrogen, under the same conditions, not necessarily N.T.P. Absolute density is thus to be expressed as a mass, and relative density as a pure number.

After the general acceptance, about the middle of last century, of Avogadro's theory, a rapidly increasing importance attached to the determination of gaseous and vapour densities as a means of establishing molecular and atomic weights. Methods of later date, especially those based on the principle of limiting densities, must be ranked among the most exact means at the chemist's disposal for the evaluation of atomic weights, at least of the lighter elements. In other directions it may be noted that even an approximate value of the density of a gaseous mixture may yield important information about its composition; and that we owe the discovery of the inert gases to an investigation of the density of atmospheric 'nitrogen'. The accounts by Lord Rayleigh of his researches on the densities of common gases should be consulted as still unsurpassed examples of exact work on this subject (refs. p. 12).

The following exercises illustrate the various ways of making for gases and vapours the two estimations of (1) weight, (2) volume, to which all determinations of the density of matter must reduce.

(1) The determination of molecular and atomic weights from gaseous density

(a) *The method of limiting density*

For gases of not too high molecular weight, and at pressures not greater than atmospheric, the van der Waals equation of state becomes practically exact:

$$(p + a/V^2)(V - b) = RT, \quad \text{or} \quad pV = RT + bp - a/V + ab/V^2. \quad (1)$$

The very small term ab/V^2 may be neglected, and in the small term a/V we may put $V = RT/p$; then

$$pV = RT \{1 - [a/(RT)^2 - b/RT] \, p\} = RT \, (1 - Ap),$$

where A is a specific constant at constant temperature.

Finally, $V = M/d$, where M is the gram-molecular weight, and d is the density (g./l.) at temperature T and pressure p,

$$\frac{p}{d} = \frac{RT}{M} \, (1 - Ap), \quad \text{or} \quad M = RTd \, (1 - A)_{p=1}. \qquad (2)$$

Therefore if the density d is determined at a series of diminishing pressures, and then p/d is plotted against p, the graph becomes accurately linear as p decreases, and the intercept on the p/d axis ($p=0$) is RT/M, whence M can be exactly evaluated.*

A novel procedure was used by Rayleigh (1902 and 1905 below). The apparatus is so arranged that the volume of a specimen of gas can be observed under two (low) pressures, one of which is exactly half the first. Then putting $p_2 = \frac{1}{2}p_1$ in

$$\frac{p_1 V_1 - p_2 V_2}{RT} = A \, (p_2 - p_1),$$

we find

$$\frac{V_2 - 2V_1}{RT} = A.$$

Once A has been evaluated, M may be obtained from (2) by a single density determination. $d_0 \, (1 - A)$, where d_0 is the absolute density (N.T.P.), is termed the *limiting* density.

(b) The method of comparable gas pairs

If two gases have closely similar values of the van der Waals constants a and b, i.e. are equally imperfect, they obey Avogadro's law accurately at pressures such as atmospheric where their individual imperfections are quite evident. In such cases molecular weights may be compared by a single density determination for each gas at atmospheric pressure.

---

* R must be expressed in units compatible with those used for p and d; in litre-atmospheres per gram-molecule per degree $R = 0.08205$.

From the isoelectronic pairs (i) CO_2 and N_2O, (ii) N_2 and CO, and from the pair NO and O_2, Guye succeeded in obtaining very accurate estimates of the atomic weights of nitrogen and carbon.*

(2) The absolute densities of gases

(a) *To determine the absolute density of dry air (free from carbon dioxide)*

In Regnault's original method, on which that here described is based, a glass globe of suitable capacity in relation to the density of the gas (Morley used one of 10 l. for hydrogen) was highly exhausted, weighed, filled with the gas (generally to normal pressure) at a known temperature, and then reweighed; the volume of the gas was taken as equal to the capacity of the globe, which was usually ascertained by filling with water and weighing. The chief experimental difficulties are those always to be expected in weighing glass apparatus of large size and surface: (1) the hygroscopic nature of glass, the surface of which on the molecular scale may be likened to a sponge, and so may retain by adsorption a variable amount of water (depending on the hygrometric state of the surroundings) whose weight may be an appreciable fraction of that of the contained gas; (2) *changes* of buoyancy during the experiment. Both these difficulties may be obviated simultaneously by counterpoising the filled globe by another of the same glass and nearly the same volume and surface, but it is well to

* (1) Exact determination of gas density: Lord Rayleigh, *Proc. Roy. Soc.* **50**, 448 (1892), oxygen and hydrogen; **53**, 134 (1893), air, oxygen and nitrogen; **55**, 340 (1894). Lord Rayleigh and Ramsay, *Phil. Trans. Roy. Soc.* A, **186**, 187 (1895), atmospheric argon. (2) Molecular and atomic weights from gas density: Lord Rayleigh, *Phil. Trans. Roy. Soc.* A, **198**, 417 (1902), A, **204**, 351 (1905). Guye, *J. chim. physique*, **3**, 321 (1905), **5**, 203 (1907), **6**, 769 (1908), method of limiting density and comparable gas pairs; **17**, 141 (1919), density and molecular weight of hydrogen bromide. Guye and Batuecas, *J. chim. physique*, **20**, 308 (1923), oxygen, hydrogen, and carbon dioxide.

verify the consistency of the results obtained in determinations with globes of different sizes.

In the method below the volume of the gas is measured independently in a gas burette, so that the globe need not be exhausted beyond the power of the water-pump. A stout-walled rubber connexion well wired on, with a good screw-clip (fig. 3b) is to be preferred to a glass tap, unless the barrel of this is 'vacuum' bored (fig. 3a). The globes, which may have a capacity of 200–400 ml. except for hydrogen, are more conveniently attached to the balance if they are pear-shaped. The absorption

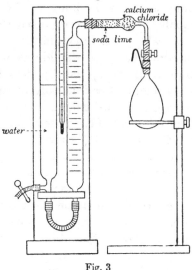

Fig. 3

tube is filled partly with soda-lime (at the end nearer to the burette) and partly with granular calcium chloride, the two reagents being separated by a small plug of glass or cotton-wool.

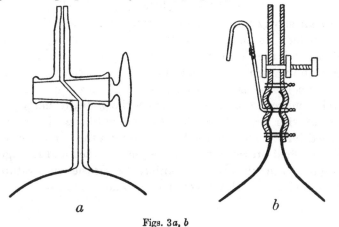

a   *b*

Figs. 3*a*, *b*

Charge the gas burette with about 230 ml. of air free from carbon dioxide by first filling with water to the capillary and then running it out slowly through the exit tube (fig. 3). The water-levels in the limbs of the burette are automatically equalized. Exhaust the globe for 2–3 min. on the water-pump (preferably with safety bottle interposed to prevent water entering the globe should the water pressure be suddenly lowered), screw up the clip, detach the globe, and *lastly* turn off the pump. Read the burette carefully with the aid of white paper inserted behind, and then test the globe for leakage by connecting it firmly to the absorption tube, without opening the clip (or tap). If the level in the burette has risen at all after 5 min. there is leakage at the clip, which should be at once adjusted, or, if necessary, the exhaustion and filling of the burette repeated. Detach, wipe over with a cloth the globe and a second similar one to serve as counterpoise. Attach the globes by their wires to the upper hooks of the balance pans, ascertaining that the rubber or glass connexion does not interfere with the action of the stirrup above. Record the weights required for balance in the form shown below. Note the temperature on a thermometer hanging on the burette at intervals during the filling of the globe as follows. After attaching the globe again to the burette, fill the ungraduated limb of the latter with water, and *then* slightly release the clip, so as to draw air slowly into the globe through the absorption tube (to prevent its sudden release owing to 'sticking' of the rubber, rock the clip gently to and fro during unscrewing); add water to the ungraduated side as required. Screw up the clip when about the 20 ml. level is reached, carefully equalize levels, and read the burette. Detach the globe, recharge the burette with air, and proceed as above until the clip can be opened fully. Equalize the levels and read the barometer to give the final pressure in the globe; now detach and reweigh the globe, using the original counterpoise. If the temperature has varied appreciably take a mean value.

Example

Total volume withdrawn from burette = 257 ml.
Mean temp. = 14·0° (variation 0·5°). Aqueous tension = 11·9 mm.
Barometer height = 765 mm.

Weighings:

Left pan	Right pan	Difference
Exhausted globe	Counterpoise	
1·310	0·0036	1·3064 g.
Filled globe		
1·000	0·0070	0·9930 g.

Weight of air = 0·3134 g.

Volume of air in the globe *at* 765 *mm.* $= 257 \times \dfrac{765 - 11·9}{765}$.

Volume (V_0) reduced to N.T.P. $= \left[257 \times \dfrac{753}{765} \right] \times \dfrac{765}{760} \times \dfrac{273}{287}$ ml.

Density $= \dfrac{0·3134}{V_0} \times 1000 = 1·29(6)$ g.

To adapt the apparatus for gases other than air insert
between the absorption tube and the globe a T-joint, through
which the globe is exhausted and gas delivered to the burette:
with this arrangement it is not necessary to detach the globe
during the filling.

(b) *To determine the absolute density of oxygen*

The serious difficulties encountered in Regnault's method
need not arise when the gas can be completely absorbed by,
or readily generated from, some reagent of small bulk. As
a generating substance should contain a high proportion of
available gas, liberated in a pure state on gentle heating, and
leaving products differing as little as possible in bulk from the
undecomposed material, a choice is not always easy. For
oxygen, potassium permanganate meets the requirements
adequately $(4KMnO_4 = 2K_2O + 4MnO_2 + 3O_2)$; spongy palla-
dium, which had been allowed to 'sorb' hydrogen, was
employed by Morley as generator of that gas.

The gas burette of exp. *a* is again required, but the absorption tube should be filled wholly with calcium chloride. The bent end of the tube is connected through a short piece of stout-walled rubber tubing, with screw-clip, to a bent tube *b* (fig. 4), which carries a rubber stopper. Choose a dry, thin-walled test-tube into which the stopper fits well, and place in it about 2·5 g. of pure powdered potassium permanganate. About 1 cm. above the powder put a loose plug of glass (*not* cotton) wool, and then a small quantity of granular calcium chloride, kept in place by a second glass-wool plug (fig. 4). Close the tube with a well-fitting solid rubber stopper, hang it by a wire from the hook of the balance pan, and weigh.

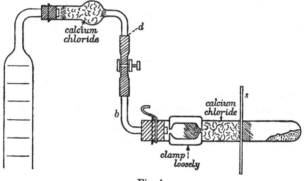

Fig. 4

Fill the gas burette with water to about the 10 ml. mark. On removing the solid stopper, attach the weighed test-tube immediately to the other, already in position on the apparatus. The screw-clip is to be kept open until the close of the experiment. Run out nearly all the water from the ungraduated limb of the burette, and observe whether a continued fall of level in the other limb indicates a leakage requiring adjustment. If none is detected, equalize the levels, and read the burette. Set a shield of asbestos paper on the test-tube in the position shown in the figure, and proceed to heat the permanganate *very* gently with a small flame (a temperature of only 150–200° is required). The manganese dioxide produced (equation above) is a very fine powder which may be carried in the gas stream out of the weighed tube if the decomposition is too rapidly effected. Although potassium permanganate is anhydrous and no water is formed in the reaction, fine powders retain surface films of water, which in this case is held by the calcium chloride; the shield is intended to prevent this from being heated.

Cease heating when about 150 ml. of gas have been collected. When the apparatus has regained room temperature (denoted by no further decrease of volume), read the burette after equalizing levels, note the temperature and the barometer height. Screw up the clip, remove the test-tube with exit tube and connexion by detaching at d (fig. 4). Connect to the water-pump through a spare calcium chloride tube, open the clip, and exhaust the tube; allow dry air to fill the test-tube through the spare drying tube; quickly replace the solid stopper and weigh the tube.

Example

Loss of weight of generating tube $= 0\cdot189$ g.
Volume of gas collected at $15°$ and 764 mm. $= 141$ ml.
Aqueous tension at $15° = 12\cdot7$ mm.
Net gas pressure $= 751\cdot3$ mm.

Volume (V_0) reduced to N.T.P. $= 141 \times \dfrac{273}{288} \times \dfrac{751\cdot3}{760}$ ml.

Absolute density $= \dfrac{0\cdot189}{V_0} \times 1000 = 1\cdot43(2)$ g.

(c) *To determine the absolute density of carbon dioxide*

When a gas possesses a definite acid or alkaline function (e.g. carbon dioxide, hydrogen chloride or ammonia), a separate determination of its equivalent weight may be substituted for direct weighing in the density measurement. The loss of weight when a specimen of pure calcium carbonate, such as calcite, is ignited to constant weight is first determined, and then the weight of hydrogen chloride neutralized by the resulting lime. We find, in an experiment which can be made with high accuracy and *without the assumption of any formulae*, that 1 g.

of hydrogen chloride is equivalent to 0·6027 g. of carbon dioxide.

The method given below has also the advantage that the density of the acid or alkaline gas can be found even when it is mixed with other gases, provided these are neutral and not appreciably soluble in the absorbing reagent.

The necessary apparatus is shown in fig. 5. The flask A carries a well-fitting rubber stopper with inlet tube, and may be of 250 ml. capacity, while B should be larger (about 300 ml.). Charge the flask A with dry carbon dioxide, either by streaming displacement, or better by previous exhaustion on the water-pump. If the gas is produced in a Kipp's apparatus it should pass through a wash-bottle to remove hydrogen chloride. Close the screwclip c and attach the bent capillary tube t (about 1 mm. bore), which dips to the bottom

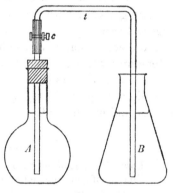

Fig. 5

of flask B containing about 200 ml. of an approximately $N/5$ solution of sodium hydroxide. Open the clip, and immerse flask A for a few minutes in ice-water and gently shake, to start the intake of the solution. When about 20 ml. have passed in, remove the flask from the cooling bath, but continue shaking until no more solution enters the flask. A good supply of alkali should be maintained in B, and on no account must the end of the bent tube be allowed at any time to emerge from the liquid. The final level of the liquid in A will probably reach the neck of the flask.

Equalize the levels in A and B by adjusting the amount of solution in B, and extract the bent tube from the rubber connexion. Then remove the stopper and tube, take the temperature of the solution in A, and read the barometer. Any volume of the solution now extracted from A contains, as freshly dissolved carbonate, an exactly equal volume of carbon dioxide measured under the conditions of temperature and pressure just observed.

To 50 ml. of solution from A add about 20 ml. of reagent barium chloride and titrate with standard (about $N/4$) hydrochloric acid, using *phenolphthalein*; only the excess sodium hydroxide present will be titrated. Repeat the titration, again using barium chloride and phenolphthalein, with 50 ml. of unused alkali solution, which will undoubtedly also contain carbonate. The difference between the titrations clearly represents the hydrochloric acid equivalent to the 50 ml. of gas dissolved in 50 ml. of alkali during the experiment.

Example

Barometer height = 776 mm. Temperature of final solution = 18·5°. Hydrochloric acid = 9·124 g./l.

Difference between titrations of 50 ml. of sodium hydroxide solution before and after absorption = 17·3 ml.

Assuming that 1 g. of hydrogen chloride is equivalent to 0·6027 g. of carbon dioxide (see above), weight of 50 ml. of carbon dioxide, at 776 mm. and $18·5° = 17·3 \times \dfrac{9·124}{1000} \times 0·6027 = 0·0951$ g.

Absolute density of carbon dioxide = 1·99 g.

(3) The determination of vapour density

As the absolute density of a vapour will usually have no direct physical meaning, the term 'vapour density' is taken to mean relative density. In calculations the absolute density of hydrogen (0·0899 to three figures) may be taken as 0·09.

(a) The method of Dumas

This method, which, except for the manner of filling the bulb, is closely similar to that of Regnault (p. 12), has proved especially suitable for the study of dissociating vapours, as it yields accurate results, and the temperature of the determination can be readily varied.

Boil about 500 ml. of water in a flask for 10 min. to expel dissolved air, and then allow it to cool undisturbed. Select a

bulb with a long projection, support it on the balance pan
with a crucible and weigh. Insert the projection into a test-
tube supported in the rack, and containing not less than 10 ml.
of the liquid to be used. By alternate warming and cooling
draw the liquid into the bulb, which is then placed in a wire
holder (fig. 6), and immersed as completely as possible in a
water-bath; this may be previously heated to about 50° but

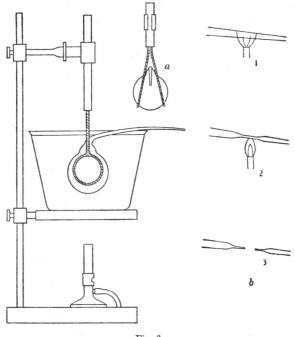

Fig. 6

should not be boiling. The projection should, if possible, point
downwards. Raise the temperature of the bath to boiling-
point, and keep at this temperature for at least 10 min. Read
the barometer, and ascertain the temperature of the bath from
tables of the boiling-point of water (or this may be computed
on the assumption that the boiling-point changes 1° for a
change of 27 mm.).

While the bath is still vigorously boiling, and without moving the bulb, seal the narrow end of the projection in the manner shown in fig. 6 b, retaining the drawn-off portion undamaged (the sealing must be done by previously drawing out, as indicated, and not by merely fusing over the extreme end of the tube into a blob of glass that invariably cracks and admits air as the bulb cools). A small spirit lamp is very convenient for this purpose. Inspect the seal through a hand lens, and if it is found satisfactory withdraw the bulb, wipe it over, and weigh together with the portion drawn off in sealing.

Place at least 300 ml. of the now cool air-free water in a porcelain dish, insert the sealed end of the bulb under the water, and break the end by pressure between the jaws of crucible tongs (*not* by jabbing on the bottom of the dish). Water should immediately enter and very soon nearly fill the bulb; if the liquid used is insoluble in water (e.g. chloroform), a small bubble composed of liquid and saturated vapour must remain, but this should not exceed 1 ml. in volume. Much air will accumulate in the bulb, and the experiment be spoiled, if water containing dissolved air is used. Weigh the filled bulb on a 'rough' balance to the nearest 0·1 g.

Example (chloroform)

Weight of empty bulb = 22·426 g.
Weight of bulb, sealed, with vapour = 22·730 g.
Difference = 0·304 g.
Boiling-point of water at 767 mm. = 100·3°.
Weight of bulb filled with water = 133·1 g.
Capacity of bulb = 133·1 − 22·4 = 110·7 ml.
Buoyancy of sealed bulb = volume × density of air, at pressure and temperature of balance room (18°)

$$= 0.1107 \times 1.29 \times \frac{273}{291} \times \frac{767}{760} = 0.135 \text{ g.}$$

True weight of vapour = 0·304 + 0·135 = 0·439 g.
Weight of 110·7 ml. of hydrogen at 100·3° and 767 mm.

$$= 0.1107 \times 0.09 \times \frac{273}{373} \times \frac{767}{760} = 0.00736 \text{ g.}$$

Relative density of chloroform vapour $= \dfrac{0.439}{0.00736} = 59.6.$

(b) The method of Victor Meyer*

In the original form of this famous method the volume of vapour produced from a weighed small quantity of liquid is found by causing it to displace an equal volume of air from a special apparatus. The method was first devised to supplement Dumas' method for difficultly volatile substances requiring a high temperature bath, such as boiling sulphur (444°) or molten lead. The procedure has the great advantage that the temperature of the bath does not enter into the calculations, and therefore need not be accurately known, but it should always be at least 50° above the boiling-point of the liquid vaporized.

Suitable dimensions for the vaporizing part of the apparatus are shown in fig. 7. The bath may be of aluminium† in place of the more fragile glass of the classical design. The expelled air should be collected in a gas burette (Hempel), or failing this, in an ordinary burette inverted, with rubber stopper and levelling funnel as shown in the figure. Remove the stopper s, fill the bath to about two-thirds of its capacity with water, which is then kept at boiling temperature while the specimen of liquid is weighed. Weigh a small drawn-off tube (fig. 7, t), and then introduce into it by alternate warming and cooling about 0·1 g. of the liquid; lightly seal over the end, allow to cool and reweigh. The temperature should by now be steady in the apparatus; to test this, open the tap of the burette and fill with water to the 50 ml. mark, afterwards clamping the funnel in a fixed position; the levels of water in burette and funnel are automatically equalized. Insert the stopper, when the level in the burette will at once be slightly depressed by the air displaced by the stopper, but will not continue to fall if the temperature is steady in the bath (fluctuations through about 0·2 ml. are generally unavoidable). Equalize the levels and

* See Victor Meyer, *Abstr. Chem. Soc.* **44**, 618 (1883), or *Ber.* **15**, 2775 (1882).

† A serviceable bath can be cheaply obtained by cutting out the screw-top from a circular aluminium 'foot-warmer'.

read the burette. It is from this reading, and not from 50 ml.,
that the volume of expelled air is to be finally reckoned.
Remove the stopper and set the water in the burette again to
the 50 ml. mark. Break off the end of the weighed tube, at
once drop it into the heated bulb and quickly replace the

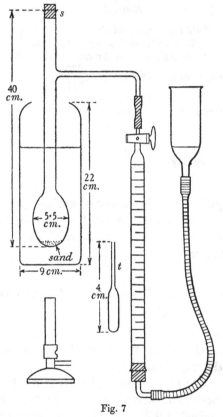

Fig. 7

stopper. As the air accumulates in the burette maintain the
pressure near atmospheric by lowering the funnel.

    When the water-level ceases to fall but only fluctuates
slightly, turn off the tap of the burette, detach from the
vaporizing bulb, and remove to a place of fairly constant temperature, which is noted on a thermometer hanging on the

burette. When the air has assumed this temperature, read the level after equalizing, and also the barometer. To prepare the apparatus for a second determination suck out the vapour by a bent tube attached to the water-pump.

Example

Weight of chloroform $= 0.2312$ g.
Air collected, at $18°$ and 773 mm. $= 45.4$ ml.
Aqueous tension at $18° = 15.4$ mm.

Corrected volume $= 45.4 \times \dfrac{757.6}{760} \times \dfrac{273}{291}$

$\qquad = 42.4$ ml. at N.T.P.

Weight of 42.4 ml. of hydrogen at N.T.P.

$\qquad = 42.4 \times \dfrac{0.09}{1000} = 0.00382$ g.

Relative density $= \dfrac{0.2312}{0.00382} = 60.5.$

Suitable liquids for use in the above methods (*a*) and (*b*) are: chloroform, acetone, methyl alcohol, methyl acetate. Care should be exercised in the Dumas method when the issuing vapour is inflammable.

(c) *Analysis of binary liquid mixtures by determination of vapour density*

When the molecular weights of the constituents of a binary mixture differ sufficiently, the composition may be found by determining the vapour density.

The constituents of a mixture of methyl alcohol ($M = 32$, boiling-point $64.7°$) and methyl acetate ($M = 74$, boiling-point $56°$) are both miscible with water; the densities of the pure liquids are not very different (0.796 and 0.940), and the densities of mixtures of the two liquids deviate widely from the simple mixture law; the two liquids form a constant-boiling mixture (see p. 83). In such a case a determination of vapour density is much the simplest physical method of finding the composition of a mixture.

Distil slowly from a small distilling flask fitted to a water condenser a mixture of methyl alcohol and methyl acetate in the volume proportions 1:4 respectively. Record the (con-

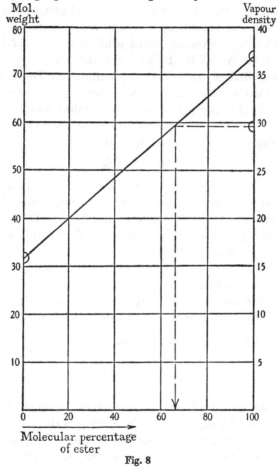

Fig. 8

stant) boiling-point (about 54°). Determine the vapour density of the distillate by Dumas' or V. Meyer's method.* Find graphically the composition of the distillate (the constant-boiling mixture), as indicated in fig. 8. (The mixture should be found to contain 18 % by weight of alcohol.)

* For the latter method use not more than 0·1 gm. of liquid.

(4) Gaseous diffusion

If we investigate experimentally the relationship of pressure and volume, at fixed temperature and not too high pressures, for specimens of various gases whose volumes are equal (say 1 l.) under some fixed standard conditions (say N.T.P.), we shall find the law of Boyle, $pV = $ constant, to reproduce the experimental data to a close approximation. The constant will assume the same value for all gases.

Adopting the kinetic theory in its simplest form, in which molecules are regarded as point-masses, we may readily calculate an expression for the product pV:

$$pV = \tfrac{1}{3}Nm\bar{c}^2 = \tfrac{1}{3}d_0\bar{c}^2,$$

where N is the number of molecules, each of mass m, contained in the volume V, and $\bar{c}$ is the 'average' velocity (actually the root-mean-square velocity) of a molecule; d_0 is the absolute density of a gas at N.T.P., since it is the fixed total weight of gas in the volume V, which is 1 l. at N.T.P.

Combining the results of experiment and calculation, we have $d_0\bar{c}^2 = $ constant, or in words—the average velocity of a gas molecule at constant temperature is inversely proportional to the square root of the absolute density at N.T.P.

Further, since at constant temperature $d_p = d_0 \times \dfrac{p}{760}$,

$$\sqrt{\frac{p}{d_p}} = \text{(constant)}\ \bar{c}. \tag{1}$$

It may also be shown from the kinetic theory that the number of molecules striking 1 cm.^{2} of the surface of the container of the gas is proportional to $\bar{c}$, and therefore the rate at which molecules escape into a vacuum through an orifice of molecular dimensions in the wall of the container, i.e. the rate of effusion, is given by

$$\sqrt{\frac{p}{d_p}} = k\ \text{(rate of diffusion)}. \tag{2}$$

Graham, who made a detailed study of the passage of gases

through various porous materials,* found that several types of septa could be distinguished: (1) those that behaved as assemblages of effusion plates, with holes of molecular dimensions, e.g. compressed graphite, unglazed porcelain; (2) septa with relatively large interstices, e.g. most compressed powders; (3) septa containing principally fine capillaries of great length relative to their width, e.g. plaster of Paris. Only septa of the first class are associated with the simple law (2) above.

The term 'diffusion' is now usually applied in a very general sense to passage through any type of membrane or septum, and also to the interdiffusion of one gas into another.

(a) A septum of compressed graphite, similar to that recommended by Graham, may be easily constructed as follows. The bore (2 mm.) of a thick-walled glass tube T (fig. 9) about 16 cm. long is constricted near the centre of the tube by fusing in the blow-pipe. A very small piece of white asbestos (such as is used for filtration in the Gooch crucible) is firmly pressed into the top of the constriction at a, and then powdered graphite introduced above and well compacted by firmly pressing with a square-ended wire which closely fits the bore of the tube. A plug about 7 mm. long is so formed and kept in place by inserting a second small piece of asbestos. The prepared tube is connected, with the septum uppermost, through calcium chloride tubes to the gas burette B (with water as liquid), and below to a water-pump. (The gas burette used in the V. Meyer method, p. 22, will be also suitable for this experiment.) Screw-clips, S_1 and S_2, are placed on the stout rubber connexions of the diffusion tube.

The diffusion meter so assembled is now calibrated by finding the rate of diffusion of dry air under a pressure of about 1 atm. With the clips S_1 and S_2 closed, draw air into the burette through the T-piece t, equalize the levels, connect the water-pump, and open first S_2 and then S_1. Allow about 2 ml. of air to pass the septum, and then take the time for the diffusion of exactly 5 ml. The funnel of the burette need not be moved. A second confirmatory timing may be made as the

* *Phil. Trans. Roy. Soc.* A, **153**, 385 (1863).

meniscus passes slowly up the burette. When the observations are finished, screw up S_2, detach and *then* turn off, the pump; finally, screw up S_1 and detach the diffusion tube. If air is suddenly allowed to enter the evacuated diffusion tube below the septum, the latter may be displaced and spoiled.

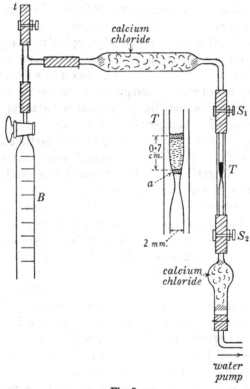

Fig. 9

(b) *Determine the specific gravity of coal-gas.* Fill the burette with water to the tap, which is then shut. Connect the gas supply to the end of the upper calcium chloride tube and thoroughly flush out this tube, allowing the gas to pass out through t into an aspirator. Then close t, and take a supply of gas into the burette to the same level as in the calibration.

Observe the time of diffusion of 5 ml. under the same condition as before. Calculate the specific gravity of the gas (air = 1) from the formula

$$d = \frac{(\text{time for coal-gas})^2}{(\text{time for dry air})^2}.$$

(c) *Determine the amount of impurity (assumed to be air) in hydrogen supplied by Kipp's apparatus.* Flush out the apparatus with the gas as in (b) above, and proceed to time the diffusion of 5 ml.

Examples

Times of diffusion of 5 ml. of gas under the same pressure difference:

Air = 477 sec.

Coal-gas (Cambridge Town supply) = 322 sec.

Gas from Kipp's hydrogen generator (with pure zinc and dilute sulphuric acid) = 160 sec.

Specific gravity of coal-gas = 0·455 (air = 1).

Relative density of gas from Kipp's apparatus (air = 14·4) = 1·60.

Volume percentage of impurity (air) = $100x$.

Then $(1-x)+14\cdot4x=1\cdot6$, whence $100x=4\cdot5$.

Chapter II

CRYSTALLIZATION AND THE PROPERTIES OF CRYSTALS

Introduction. Owing to the relatively great tenuity of the gaseous state, it has proved feasible to construct a fairly exact theory of gases without attributing to a molecule a shape other than that of a sphere. In the crystalline state the molecules are in the opposite extreme condition of such close packing that their shape, size and electric charge (if any) become all-important, and must be the paramount influences on the external form of the crystal.

It was long accepted by crystallographers as a working hypothesis that the most obvious and the most characteristic property of crystals, namely, their perfectly regular geometrical form and symmetry, was a necessary consequence of a regular arrangement of their component molecules (or ions), although this was not directly proved by experiment until the modern discovery of the method of X-ray analysis. Present-day treatment of crystal structure is based on the conception of the unit cell, first imagined by 'the father of crystallography', Haüy. The unit cell of a crystal may be described as the smallest molecular unit, by the multiplication and orderly juxtaposition of which the homogeneous crystalline structure may be indefinitely extended. The cell may, and commonly does, contain more than one molecule. That the presence of more than one molecule in a cell may result in a gain of symmetry is easily seen by considering the symmetry of the figure Σ and its reduplicated form $\Sigma\Xi$.

The crystal systems and their unit cells

The geometrical study of the external forms of crystals enabled them to be classified at an early stage into seven principal systems, according to their symmetry properties. More recently the methods of X-ray analysis have enabled the underlying unit cells to be identified, and their dimensions and characteristic angles accurately measured. It is obvious that space may be exactly filled without voids by the indefinite multiplication and juxtaposition of the unit cell in the form of

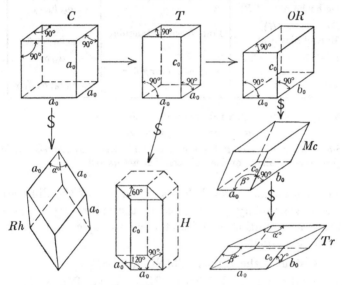

Fig. 10. *Unit cells of the seven crystal systems.* [*C*, cubic; *T*, tetragonal; *OR*, orthorhombic; *Rh*, rhombohedral; *H*, hexagonal; *Mc*, monoclinic; *Tr*, triclinic.]

a parallelepiped, with three pairs of parallel and equal faces. Fig. 10 shows how the various types of cell may be regarded as interrelated by appropriate modifications of edge lengths or by 'skewing' (indicated in the figure by the sign $). Table 1 gives a grouping according to the number of right angles, and shows the minimum number of data essential to define the cell fully. (Lengths a_0, b_0, c_0 are always expressed in Ångström units of 10^{-8} cm.)

Table 1

System	Corresponding cell	Characteristic data
Cubic (C) Tetragonal (T) Orthorhombic (OR)	Wholly rectangular	a_0 a_0, c_0 a_0, b_0, c_0
Hexagonal (H) Monoclinic (Mc)	Four faces rectangular	a_0, c_0 $a_0, b_0, c_0; \beta$
Rhombohedral (Rh) (=trigonal) Triclinic (Tr)	No face rectangular	a_0, α $a_0, b_0, c_0; \alpha, \beta, \gamma$

Axial ratios are expressed as, for example,

$$a : b : c = x : 1 : y = a_0 : b_0 : c_0.$$

Such ratios have long been in use, and with the characteristic angles give the shape of the unit cell, *but not its size*.

The compounds given below as examples of the different systems are usually to be found in the laboratory in good crystalline condition suitable for microscopic and goniometrical examination:

Cubic system: the alums, $Pb(NO_3)_2$, CaF_2.

Tetragonal system: HgI_2(red), PH_4I, $K_4Fe(CN)_6.3H_2O$.

Orthorhombic system: $ZnSO_4.7H_2O$, K_2SO_4, KNO_3.

Monoclinic system: $FeSO_4.7H_2O$, $FeSO_4.(NH_4)_2SO_4.6H_2O$, $Na_2SO_4.10H_2O$, $Na_2SO_3.7H_2O$, $Na_2S_2O_3.5H_2O$, $KClO_3$, $K_3Fe(CN)_6$.

Triclinic system: $CuSO_4.5H_2O$, $K_2Cr_2O_7$.

Rhombohedral system: $CaCO_3$ (calcite), $NaNO_3$.

Hexagonal system: SiO_2 (quartz), ice, PbI_2.

The formation of a crystal by an assemblage of unit cells is illustrated in figs. 11 and 11a, which show a drawing of a crystal of the Epsom salt type, belonging to the orthorhombic system, and a section $ABCD$ across the prismatic part of the crystal orthogonal to its long axis. To build up a crystal we first construct two axes a and b at right angles, and imagine another, the c axis, set at right angles to the plane of the paper. In general the axes to be chosen for a particular crystalline system take the direction of the edges of the appropriate unit cell—for a monoclinic crystal the c axis will be inclined at an angle β to the axis a and to the plane of the paper and the two rectangular axes a and b. We begin to build up the cells by first placing an equal number (five are shown, enormously magnified, in fig. 11a) along the axes a and b, setting the sides of length a_0 against the a axis, and the longer sides of length b_0 to the b axis. Reckoning in terms of the characteristic lengths a_0 and b_0 respectively, we have now *equal*

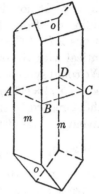

Fig. 11
Epsomite: $MgSO_4.7H_2O.$
$m:m = 89° 26'.$
$o:m = 50° 49'.$

intercepts on the two axes. More unit cells are now blocked in, as shown in fig. 11a, in the outward direction. Any lines now drawn, such as AB or AS, which bound an integral

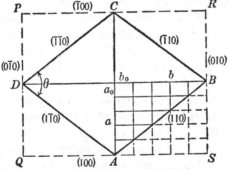

Fig. 11a. Forms of an orthorhombic crystal

$$2\tan^{-1}\frac{a_0}{b_0} = \theta.$$

number of cells represent sections (in the plane of the paper) of possible crystal faces. If in building up succeeding layers of cells above that already formed we retain always equal numbers in each layer, it is evident that the faces will grow parallel to the c axis; in other words, the intercept on this axis is ∞. It now becomes obvious that a similar process of building out from the axes in the other quadrants will yield sections AD, DC and CB, all three of which are physically identical with the original AB.

We could symbolize this set of faces by an expression of the type $\{a, b, c\} = \{11\infty\}$, embodying the intercepts on the axes without regard to sign. It has, however, become customary to avoid the use of the infinity sign by stating reciprocals of the intercepts, $\left\{\dfrac{1}{a}, \dfrac{1}{b}, \dfrac{1}{c}\right\} = \{h, k, l\} = \{110\}$. This type of expression, known after its inventor as the Millerian notation, is in universal use among crystallographers; the three integers in the bracket are known as indices. The pair of faces PQ and RS $\{010\}$ differ from $\{110\}$ but are themselves identical, because they are both formed by the juxtaposition of end-faces of the unit cell. The pair of faces PR and QS $\{100\}$, although also physically indistinguishable, differ again from PR and QS because they are formed from the $side$-faces of the unit cells. Thus our figure exhibits altogether $three$ sets of structurally different faces. Etch figures, produced by the application of small quantities of a solvent to the faces, can often bring such differences to light, and when discovered, they play a very important part in assigning a crystal to its system. A group of physically identical faces such as $\{110\} = (110)$, $(\overline{1}\overline{1}0)$, $(\overline{1}10)$, $(1\overline{1}0)$ is known by crystallographers as a $form$. Ordinary specimens of the epsomite group of hydrated sulphates show a predominant if not exclusive development of the form $\{110\}$, but the axial ratio $a:b$ is so nearly unity that the interfacial angle $(\theta = 2 \tan^{-1} a_0/b_0)$ differs very little from a right angle, and the crystals may be termed pseudo-tetragonal (table 6). The foregoing discussion explains what, at first sight, may seem to appear contradictory, namely, that a wholly rectangular unit cell can give rise to a crystal devoid of right angles.

The form $\{110\}$ alone cannot give rise to a solid figure, as there is no provision for termination of the prism. In epsomite

and analogous salts termination is often effected by the so-
called sphenoidal faces (labelled o in fig. 11), from σφήν,
a wedge. The characteristic opposite orientation of the wedges
at each end gives to these crystals an enantiomorphic nature,
i.e. its mirror image is not superposable on the crystal, and
dextro (d-) and laevo (l-) types are found. The crystals exhibit
a very feeble optical rotation. Inoculation of a supersaturated
solution of epsomite with d- or l-tartaric acid is said to promote the precipitation of the corresponding d- or l-crystals of
epsomite.*

The forms discussed above in connexion with epsomite do
not, of course, represent more than a small fraction of those
met with on crystals. The form {111} is of frequent occurrence,
and constitutes the eight identical faces of the regular octa-
hedron of the (cubic) crystal of common alum, and of many
other less regular octahedra belonging to other systems. In
general, a crystal produced without special precautions exhibits a number of forms in combination; the appearance of
the crystal is governed by the relative development of its
forms, and this is known as its *habit*.

Crystalline form and chemical constitution

Much confusion has been introduced into this subject by
very diverse interpretations of the meaning of the term
'isomorphism'. When Berzelius first suggested the use of this
term to his pupil Mitscherlich (1819), the crude goniometers of
the day did not permit of measurement sufficiently refined to
distinguish between the angles of closely similar crystals, and
Mitscherlich probably believed then that the angles between
principal faces, and the axial ratios, of the phosphates and
arsenates which he had recently examined were equal.
Measurements with modern apparatus have disclosed that
(excepting crystals of the cubic system) no two crystals of
different substances possess exactly equal characteristic data,
thus finally proving the tenet of specific crystalline shape so

* P. Groth, *Chemische Krystallographie*, 2, 429.

strongly championed by Haüy. The term 'homomorphic' would now be preferable, if the familiar 'isomorphic' could be displaced (tables 2 and 3).

Now that the actual dimensions of unit cells and not merely the axial (or edge) ratios can be ascertained and compared, it becomes necessary to introduce further descriptive terms, if confusion is to be avoided in the future: (1) molecules (or ion-

Table 2. *Homomorphism: actual cell dimensions*

(Unless otherwise stated, all data are taken from the tables in *The Structure of Crystals*, Wyckoff, Chem. Catalog Co. 1931.)

Group	Compound	Crystal system	Characteristic data			
			a_0	b_0	c_0	Angles
A	$(NH_4)_2SO_4$	Orthorhombic	5·95	10·56	7·72	
	K_2CrO_4		5·88	10·30	7·45	
	$(K_2Cr_2O_7$	Monoclinic	7·47	7·35	12·97	$\beta = 91°\,55')^*$
B	$MgSO_4.7H_2O$	Orthorhombic	11·91	12·02	6·87	
	$NiSO_4.7H_2O$		11·86	12·08	6·81	
	$ZnSO_4.7H_2O$		11·85	12·09	6·83	
C	$FeSO_4.7H_2O$	Monoclinic	15·34	20·02	12·98	$\beta = 104°\,15'$
	$CoSO_4.7H_2O$		15·45	20·04	13·08	$\beta = 104°\,40'$

For groups B and C see also table 6.

* See exp. *a*, p. 45.

pairs) may be called *homotectonic* ($\tau\acute{\epsilon}\chi\tau\omega\nu$, a sculptor) when they have closely similar shape, or, in the case of ion-pairs, nearly equal $r_{\text{kation}}/r_{\text{anion}}$ ratios. Such molecules will commonly aggregate in cells also of similar shape. (2) If, in addition to the similarity of shape, the absolute sizes (molecular volumes) are nearly equal, the molecules may be spoken of as *homoplastic* ($\pi\lambda\alpha\sigma\tau\acute{o}s$, moulded); such molecules will give cells of similar shape *and size*. Finally (3) in *homochemical* molecules the corresponding elements belong to the same periodic group, or at least have the same valency, and the bond diagrams are identical. The classical term 'isomorphic' (or

Table 3. *Homomorphism of the carbonates* of
divalent metals*

System—rhombohedral: data $= a_0$ and $\beta =$ obtuse angle of cell.

A.
$$\text{MgCO}_3 \begin{cases} 5\text{·}899 \\ 103°\ 19' \end{cases}$$

$$\begin{matrix} \text{CaMg(CO}_3)_2 \\ \text{dolomite} \end{matrix} \begin{cases} 6\text{·}14 \\ 102°\ 52' \end{cases}$$

$$\text{CaCO}_3 \begin{cases} 6\text{·}36 \\ 101°\ 55' \end{cases} \qquad\qquad \text{MnCO}_3 \begin{cases} 6\text{·}06 \\ 102°\ 50' \end{cases} \quad \text{FeCO}_3 \begin{cases} 6\text{·}03 \\ 103°\ 4' \end{cases}$$

$$\text{ZnCO}_3 \begin{cases} 5\text{·}928 \\ 103°\ 28' \end{cases}$$

$$\text{CdCO}_3 \begin{cases} 6\text{·}28 \\ 102°\ 48' \end{cases}$$

B.†

Homomorphic mixtures—(Ca, Mn) CO_3 [calcite-rhodochrosite]

Percentage MnCO₃	0·0	1·09	7·00	15·40	32·34	42·17	95·7
Typical spacing on X-ray diagram	3·075	3·055	3·020	3·005	2·975	2·948	2·850

* General references for carbonates: Brentano and Adamson, *Phil. Mag.* **7**, 513 (1929); Levi and Ferrari, *Atti. R. Accad. Lincei* (V), **33**, 516 (1924); Zachariasen, *Chem. Abstr.* **23**, 1790 (1929).

† Krieger, *Amer. Min.* **15**, 23 (1930).

better, 'homomorphic') should now be strictly reserved to describe similarity in external crystalline form, which was actually its sole original meaning.

Homomorphism is based upon homotectony in the molecules forming the crystal, but these need not be either homoplastic or homochemical. Free replacement, in all proportions, of one unit cell by another composed of different molecules ('isomorphic mixture'), or the formation of oriented overgrowths of one crystal upon another of different substance, were apt to be taken as important tests of homochemical relationship in the molecules concerned. The implied restriction proves to be too narrow, and the tests are in reality criteria of homoplasy. Tables 4 and 5 show clearly that homoplasy does not by any means necessarily imply a homochemical nature, nor, conversely, does the latter character imply homoplasy. To exhibit the fallacy of assuming such implications it need only be pointed out that from the data in table 4 equal valencies would have to be ascribed to the

Table 4. *Homoplasy in heterochemical substances*

Cubic system

	a_0
KBr	6·57
KCN	6·55

Hexagonal system

H_2O (ice) $a_0 = 4·53$, $c_0 = 7·41$ (four molecules to the cell).
NH_4F $a_0 = 4·39$, $c_0 = 7·02$ (two molecules to the unit cell).

Orthorhombic system

	a_0	b_0	c_0
$KClO_4$	8·85	5·65	7·23
$BaSO_4$	8·89	5·44	7·17
KBF_4	7·84	5·68	7·38
$BaBeF_4$	(closely similar to $BaSO_4$)		
KNO_3	5·40	9·14	6·41
$BaCO_3$	5·29	8·88	6·41

Rhombohedral system

$NaNO_3$	$a_0 = 6·32$,	$\beta = 102°\ 42'$
$CaCO_3$	$a_0 = 6·36$,	$\beta = 101°\ 55'$

References. KCN, KBr, $KClO_4$, KNO_3, $BaCO_3$, $BaSO_4$, SiO_2, $CaCO_3$, $NaNO_3$: tables in Wyckoff, loc. cit. Ice: Barnes, *Proc. Roy. Soc.* A, **125**, 670 (1929). KBF_4: *Strukturbericht*, **2**, 486. $BaBeF_4$: *Z. anorg. Chem.* **201**, 289 (1931). NH_4F: Zachariasen, *Z. physikal. Chem.* **127**, 218 (1927).

elements in the groups Cl, S, B and Be; K and Ba; O and F. Table 3 B demonstrates the characteristic gradual change of 'spacing'* between the limits set by the homoplastic (and in this case, homochemical) pure compounds in a typical 'isomorphic mixture' in all proportions.

Table 6 presents a summary of the crystallographic data for

* The spacings, obtained directly from the method of X-ray analysis, give the distances between characteristic planes of atoms or ions in the crystal. From the spacings a_0, b_0 and c_0 are calculated.

Table 5. *Homotectony in homochemical compounds*

Cubic system (NaCl type): $a_0 =$ side of unit cube $= 2$ (distance $M \rightarrow X$)

Compound	a_0	Related inert gas	r_{gas}	r_{M+}	r_{X-}	r_{M+}/r_{X-}	Type of lattice
LiH	4·10	He	0·53	0·80	1·25	0·64	NaCl
NaF	4·62	Ne	1·60	0·98	1·33	0·74	—
KCl	6·28	A	1·91	1·33	1·81	0·73	—
RbBr	6·86	Kr	1·98	1·47	1·96	0·75	—
CsI	4·55	Xe	2·18	1·67	2·19	0·75	CsCl

(In CsI, I is at the centre of the cube, surrounded by 8 Cs at the corners of the cube; distance $M \rightarrow X = \sqrt{3}/2 \, a_0$.)

No solid solutions are formed, but of course the cubic crystals exhibit perfect isomorphism.

References. LiH: Gmelin, *Handbuch*, Syst.-Nr. 20, Lithium, p. 73. Crystal constants: Bijvoet, *Rec. Trav. chim.*, **42**, 859 (1923).

the important group of the hydrated sulphates of certain divalent metals (including the 'vitriols'). In this group of substances, as in other groups of hydrates, the degree of hydration tends to diminish with rise of temperature, in a way characteristic for each hydrate, with the important result that hydrates stable at room temperature may show different degrees of hydration, e.g. $CuSO_4.5H_2O$, $MnSO_4.5H_2O$, and $MgSO_4.7H_2O$. In general the heptahydrates show the not uncommon feature of *dimorphism*, i.e. they may crystallize in either of two systems (orthorhombic or monoclinic), but the relative stability of the two forms at room temperature varies with the metal in a manner which has no obvious chemical regularity. These hydrates are thus divided into two subgroups according to which form is stable at room temperature. Within these subgroups free homoplastic mixing occurs, as it also does with the triclinic pentahydrates of Cu and Mn sulphates, and all the monoclinic hexahydrates. These last are deposited from strongly acid solutions in place of the hepta-

4-2

Table 6. *Crystallographic data on the hydrated sulphates of certain divalent metals*
(taken from P. Groth, *Chemische Krystallographie*, 2)

(For brevity, axial ratios x : 1 : y are given as x : y)	7H$_2$O orthorhombic x : 1 : y	7H$_2$O monoclinic x : 1 : y	6H$_2$O monoclinic x : 1 : y	5H$_2$O triclinic x : 1 : y
MgSO$_4$	0·9901 0·5709	1·220 1·582 β = 104° 24′	1·4039 1·6683 β = 98° 34′	0·6021 0·5605 α = 81° 30′ β = 109° 0′ γ = 104° 55′
ZnSO$_4$	0·9804 0·5631	Separated by seeding with FeSO$_4$ but too unstable to measure	1·3847 1·6758 β = 98° 12′	—
NiSO$_4$	0·9815 0·5656	—	1·3723 1·6763 β = 99° 17′ (also tetragonal)	—
CoSO$_4$	—	1·1815 1·5325 β = 104° 10′	1·3959 1·6903 β = 98° 43′	—
FeSO$_4$	Very unstable	1·1828 1·5427 β = 104° 55′	—	0·5962 0·5770 α = 81° 23′ β = 110° 28′ γ = 105° 33′
MnSO$_4$	Do.	Unstable	—	0·5893 0·5691 α = 81° 37′ β = 110° 5′ γ = 104° 59′
CuSO$_4$	—	1·1622 1·1500 β = 105° 36′ (natural boothite)	—	0·5721 0·5554 α = 82° 5′ β = 107° 8′ γ = 102° 41′

hydrates, but are not in the strict sense stable in contact with water at room temperature. When a pair of hydrates from different groups crystallize from the same solution, more complicated phenomena occur, exemplified for the case of $CuSO_4$ and $MgSO_4$ in fig. 12. From this it will be seen that $CuSO_4$ is

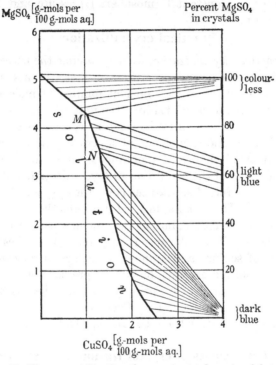

Fig. 12. The co-crystallization of magnesium and cupric sulphates.

only very slightly soluble in orthorhombic $MgSO_4.7H_2O$, and conversely, $MgSO_4$ is only sparingly soluble in triclinic $CuSO_4.5H_2O$. There is, however, a restricted region of mutual solubility round the composition about 60 % of $MgSO_4$, where both sulphates are present in the form of the monoclinic heptahydrates, forms which are unstable when the sulphates are in the pure condition. This region of solubility is made possible only by the dimorphism of the two sulphates, and the homoplasy of the (unstable) monoclinic forms. This type of relationship is often called *isodimorphism*. It will be noted that since three phases, two solid and one liquid, are present together at both the points M and N, these are non-variant points in the three-component system of the two sulphates and water, at constant (atmospheric) pressure and (room) temperature.

Practical crystallization

In practical crystallization we may assume the necessity of two probably independent processes: first, the assembling, doubtless by chance, of a sufficient number of molecules in the correct spatial relationship of the unit cell; and secondly, the growth of these (invisible) germ crystals to structures of visible size. The phenomena of supersaturation reflect the relatively small chance of spontaneous assembling, but the principal effect of the conditions of crystallization and of the presence of 'foreign' substances is on the manner of growth of the germ. Thus, the alums, with a cubic lattice, usually crystallize as octahedra, but, as Leblanc first demonstrated, a slight trace of alkali will cause the production of cubes. Pure solutions of sodium and ammonium chlorides deposit very irregularly shaped crystals, but in the presence of some urea or acetamide a crop of beautifully regular cubes with octa-hedral facets is obtained. The effect of borax on the crystal habit of Epsom salt will be seen in exp. 1 (*b*), p. 44.

(1) The process of crystallization from solution

(*a*) *Supersaturation*

In the complete absence of the solid solute, or a substance homomorphic with it, solutions can generally be obtained greatly exceeding in their proportion of solute that which under like conditions of temperature is in equilibrium with the solid, i.e. the saturated solution. Such solutions are termed *supersaturated*, and are typical examples of a metastable system. The supersaturation is at once relieved when even a trace of solid, or of a homomorph, is brought into the solution.

Solutions of gases in liquids also often exhibit a high degree of supersaturation, especially when the gas is formed in a reaction taking place in the liquid, e.g. nitrogen when liberated from a diazonium salt, oxygen from the decomposition of hydrogen peroxide.

A solution with any considerable degree of supersaturation must be used in the vessel in which it has been prepared; it cannot, of course, be filtered. Thoroughly clean three boiling tubes or test-tubes. $Na_2SO_4.10H_2O$ and water are to be placed in each tube in the proportions of 15 g. of crystals to 6 ml. of water. Put in the solid first, and run the water from a burette down the sides of the tube to wash off all solid. Close the tube with a plug of cotton-wool (not a cork) and heat without shaking in a water-bath not above 40° (to avoid the deposit of anhydrous salt). When a clear solution, *with no remaining trace of solid*, has been formed, allow the tubes to cool with the least possible disturbance.

The following substances may be used for inoculation tests: potassium sulphate (K_2SO_4), potassium chromate (K_2CrO_4), hydrated sodium chromate ($Na_2CrO_4.10H_2O$), and the original decahydrate. On relief of the supersaturation the process of crystal growth is seen at its best if the crystallization starts from the top of the liquid. To ensure this, coat the end of a glass rod with adherent crystals by moistening with a solution and allowing to dry, and insert into the tube so as just to touch the surface of the solution. The potassium salts should be once recrystallized to free from possible traces of sodium salt. If sodium chromate is not available, a specimen may be prepared by dissolving 1 g. of chromic acid in cold water, adding 1·5 g. of sodium *bicarbonate*, and allowing the solution to evaporate at room temperature (or not above 30°).

If one of the prepared tubes is left undisturbed for 48 hr. at room temperature, the metastable heptahydrate ($Na_2SO_4.7H_2O$) will usually crystallize spontaneously.

Preparations of other sodium salts may be made as follows (in all cases use scrupulously clean tubes):

(i) *Carbonate:* heat the decahydrate and water in the

proportions 10:4 respectively in a water-bath not above 50°.

(ii) *Thiosulphate:* heat 10 g. of crystals and 4 ml. of water in a boiling water-bath.

(iii) *Acetate:* heat 10 g. of crystals with 5–6 ml. of water in a boiling water-bath.

(For thiosulphate and acetate the proportions of salt to water will be sufficiently correct if the dry crystals are just covered with water before heating.)

Salt	Solubility (anhydrous salt in 100 g. water)		Hydrate deposited
	10°	20°	
Na_2SO_4	9·0	19·4	$10H_2O$ (7H_2O spontaneously on standing)
Na_2CO_3	12·6	21·4	$10H_2O$
$Na_2S_2O_3$	59·7	70·1	$5H_2O$
$NaC_2H_3O_2$	41·0	46·0	$3H_2O$

(b) The influence of foreign substances on crystal habit *

Dissolve 14 g. of Epsom salt in 20 ml. of distilled water; filter if necessary, and then divide the solution into two equal parts. Dissolve 0·1 g. of borax in 1 ml. of water and add to one part. Set aside the solutions, contained preferably in glass dishes, to evaporate slowly, during which process they must be protected from dust. Select, either by hand, or by careful filtering, well-formed crystals before the solutions have completely evaporated to dryness. Drain the crystals by turning them over and over on filter-paper, but do not wash them with water. Allow the crystals to dry in the air (on no account in the steam oven). The crystals will repay inspection under the low power of a microscope; before placing a crystal on the slide for this purpose, *very gently* roll it in the fingers to remove

* Cf. von Hauer, *Z. Kryst.* **6**, 524.

micro-crystals that will probably have formed upon it during drying.

Careful focusing will show that the crystals from the solution without borax are long, four-sided prisms, resembling the drawing in fig. 11. The sphenoidal terminal faces (*o*, fig. 11, p. 33) are usually very well developed in epsomite. At first sight the crystals from the solution containing borax seem to bear no resemblance to those from the solution of the pure salt. They appear to show a tetrahedral habit. Close inspection will, however, show that this appearance has been brought about by the almost complete suppression of the prismatic faces, leaving the sphenoidal faces dominating the shape.

(2) Isomorphism and isodimorphism

(a) *The crystallographic relationship of sulphate, chromate and dichromate* (see table 2, p. 36, group *A*)

To 100 ml. of a solution of ammonium sulphate saturated at room temperature* add 5 drops of 0·880 ammonia solution and 0·5 g. of powdered potassium chromate, which is dissolved by shaking or by gentle warming. 50 ml. of the solution *S* so prepared is placed in a boiling tube with 10 g. of solid ammonium sulphate, and the remainder set aside. The tube is heated in a boiling water-bath and its contents well stirred, preferably with a glass ring stirrer, until all the solid has dissolved. Evaporation is minimized by closing the tube with a grooved cork. Withdraw the tube and cool in running tap water with continued stirring, to promote rapid crystallization and the formation of small crystals free from inclusions. After allowing the tube and contents to stand at room temperature for about half an hour, the deposited ammonium sulphate is filtered at the pump, the mother-liquor passing into a dry receiver. Set aside this filtrate *F*, and wash the crystals with

---

* Prepare this solution beforehand by drawing a current of air by the water-pump through a mixture of 200 g. of the salt and 250 ml. of water: pour off or filter from excess solid.

a small quantity of saturated ammonium sulphate solution (not water); after well pressing on the filter, transfer the solid to filter-paper and allow it to dry in the air. Note that the crystals remain bright yellow even after washing. The chromate in 20 ml. of (i) the filtrate F, (ii) the original solution S, is estimated by treatment with acidified potassium iodide and titration of the liberated iodine with $N/10$ thiosulphate (see example below). The difference gives the amount of chromate $(CrO_4^=)$ dissolved in 10 g. of solid ammonium sulphate. This estimate may be checked by a direct analysis by the same method on the air-dried crystals. The percentages of $CrO_4^=$ in the liquid and solid solutions are found to be comparable.

Repeat the above experiment, substituting potassium dichromate for the chromate, and 2 ml. of dilute sulphuric acid for the ammonia. No difference will now be found in the estimates of chromate, and the crystals of sulphate formed are quite colourless after washing.

If the acid is omitted, the two estimates of chromate will still not differ appreciably, but the crystals will remain very faintly but persistently yellow even after washing. This observation affords a very sensitive proof of the existence of the equilibrium $Cr_2O_7^= + H_2O \rightleftharpoons 2H^+ + 2CrO_4^=$ in aqueous solutions of chromates and dichromates.

Example

Data required:

Solubility of $(NH_4)_2SO_4$ $(M = 132)$ at $15-18° = 74.5$ g./100 g. water.

Specific gravity of the saturated solution at $15-18° = 1.24$.

Wt. of $SO_4^=$ in 50 ml. of saturated solution

$$= 50 \times 1.24 \times \frac{74.5}{174.5} \times \frac{96}{132} = 19.32 \text{ g.}$$

Wt. of $SO_4^=$ in 10 g. of crystals $= 10 \times \dfrac{96}{132} = 7.27$ g.

Method of estimation: $\underset{116}{CrO_4^=} = \tfrac{3}{2}I_2 = \underset{474}{3Na_2S_2O_3}.$

Sodium thiosulphate solution contained 15.14 g. $Na_2S_2O_3$ per litre.

1 ml. of this solution $= 0.01514 \times \dfrac{116}{474} = 0.00375$ g. $CrO_4^=$.

Readings:

	Thiosulphate ml.
20 ml. of filtrate F (mother-liquor) added to 0·5 g. KI dissolved in excess (50 ml.) of dilute HCl, required	17·2
20 ml. of control (solution S) treated as above	26·9
Difference	9·7

Chromate (CrO_4^-):

In mother-liquor (50 ml.)	In crystals (10 g.)
$= 17·2 \times 0·00375 \times \frac{5}{2}$	$= 9·7 \times 0·00375 \times \frac{5}{2}$
$= 0·1593$ g.	$= 0·0898$ g.

Ratio CrO_4^- : SO_4^- :

$$\frac{0·1593}{19·32} = 0·00825 \qquad\qquad \frac{0·0898}{7·27} = 0·0124$$

(b) The isomorphism of the nickel and cobalt ammonium sulphates

Advantage may be taken of the complementary colours of nickel and cobalt salts to demonstrate the existence of mixed crystals.

To 3 g. of cobalt sulphate, 5 g. of nickel sulphate are added and both dissolved together in 10 ml. of hot water acidified with a few drops of dilute sulphuric acid. A hot solution of 5 g. of ammonium sulphate in 10 ml. of water is added and well mixed; on cooling, crystals readily separate in quantity. After filtering and washing with a small amount of water, these are found to be neutral grey in colour in daylight and quite without other tint. In electric light a green tint is visible; in green or red light the crystals appear almost black. Other proportions of the two sulphates will be found to give either pink or green mixed crystals.

(c) The isodimorphism of magnesium and copper sulphates

For discussion see p. 41. Fig. 12 shows graphically the relation between the composition of solution and that of the solid in equilibrium with the solution.

10 g. of Epsom salt and the following weights of copper sulphate pentahydrate are dissolved by boiling with 10 ml. of

water acidified with sulphuric acid (5 ml. of dilute acid to 100 ml.) in separate labelled test-tubes or boiling tubes:

Preparation no.	1	2	3	4	5
$CuSO_4 . 5H_2O$ (g.)	1·5	2·0	2·8	4·0	6·0

In each case, after obtaining a clear solution cool the tube and contents in running tap water, and stimulate crystallization by scratching the inside of the tube with the end of a glass rod. Preparations 1–3 may also be seeded with a trace of epsomite crystals. When a considerable quantity of crystals has been deposited, the tube should be closed with a plug of cotton-wool and the preparation left for some hours (preferably overnight) at room temperature, then filtered at the pump. The crystals are well pressed on the filter and then transferred *without washing* to filter-paper and allowed to dry in the air.

The specimens are found to have the following characteristics:

Nos. 1 and 2. Colourless, with the familiar needle-shape habit of Epsom salt (see p. 34 and fig. 11 for description).

Nos. 4 and 5. *Light* blue (cf. $CuSO_4 . 5H_2O$). No needles can be discerned under magnification, but only compact block-like crystals, which at first sight appear to be very 'skewed' in build. Further inspection and careful focusing will, however, disclose that the cross-section is rectangular, and the shape is that of the unit cell of the monoclinic system (fig. 10, unit cell Mc).

No. 3. If the room temperature is 16–20°, this preparation should yield colourless needles of Epsom salt (very nearly free from copper) side by side with compact light blue crystals like Nos. 4 and 5 (see diagram, fig. 12, p. 41).

If time allows, the following further preparations may be made: Dissolve in 10 ml. of hot acidified water, as in the previous preparations, the following weights of the two sulphates:

Preparation no.	6	7	8
$CuSO_4 . 5H_2O$ (g.)	7·0	7·0	6·0
$MgSO_4 . 7H_2O$ (g.)	9·0	5·0	3·0

Nos. 7 and 8. *Dark* blue tablets indistinguishable in colour and shape from pure $CuSO_4.5H_2O$. The high proportion of facets on the majority of the crystals is characteristic. The general outline is that of a very 'skewed' structure; it possesses no right angles, and not even an axis of symmetry, but only a *centre* of symmetry.

No. 6. At 16–20° this should yield a mixture of light blue and dark blue forms, but the sudden break in the character of the product is naturally not so obvious from the colour difference as in the former case (No. 3).

A parallel set of preparations in which an equal weight of nickel sulphate replaces the cupric sulphate yields a series of mixed crystals of *continuously* increasing depth of (green) colour.

Alternatively, the copper preparations may be modified as follows:

To the hot solutions of magnesium and copper sulphates add a hot solution of 5 g. ammonium sulphate in 10 ml. of water. Mixed crystals of the double ammonium sulphates of Cu and Mg readily separate on cooling, and these being 'isomorphic' (unlike the parent salts) no break in the colour series will now be detectable.

Other interesting cases of isodimorphism which may be easily investigated are to be found in the relations of $FeSO_4.7H_2O$ with either $MgSO_4.7H_2O$ or $ZnSO_4.7H_2O$.* Owing to the feeble coloration of $FeSO_4$ no striking colour differences are to be noticed in these series, but diagrams like fig. 12 may readily be constructed (if time allows) by means of simple analysis of the solutions and solids with potassium permanganate solution.

* $MgSO_4$-$FeSO_4$: Retgers, Z. *physikal. Chem.* 3, 534 (1889); Gmelin, *Handbuch,* Syst.-Nr. 27, Magnesium (B), 230.

(d) *Relationships of the cubic (perfectly isomorphic) substances
sodium chloride, chlorate and bromate*

Solubilities (g./100 g. water)	At 0°	At 20°	At 100°
NaCl	35	36	39
NaClO$_3$	82	99	233
NaBrO$_3$	27	38	91

(i) Dissolve 3 g. of sodium chloride and 7 g. of sodium bromate in 14 ml. of boiling water. Cool in running tap water, collect the crystals by filtration at the pump, wash twice with a few ml. of a saturated solution of sodium bromate (3 g. to 10 ml. of water), and dry in the steam oven. About 5 g. of product should be obtained.

Prove the absence of chloride in the following way:

Dissolve 0·2 g. in 10 ml. of water, boil, add a few drops of concentrated nitric acid, and then a solution of silver nitrate. There should be no precipitate. A white granular precipitate forming on cooling is $AgBrO_3$, which is only moderately soluble in cold water.

(ii) Dissolve 8 g. of sodium chlorate and 5 g. of sodium bromate in 12 ml. of boiling water. Proceed then as in (i); about 5 g. of dry crystals should be obtained.

Test for the presence of both chlorate and bromate as follows:

Heat 0·3 g. of the crystals cautiously in a hard-glass tube until the effervescence of oxygen has ceased; then fuse the product at red heat. After cooling, dissolve in about 20 ml. of hot water, transfer to a flask, and acidify with a few drops of concentrated nitric acid. Add 0·1 g. of potassium iodate and boil. Fumes of bromine are at once seen; continue boiling until the liquid is colourless (15–20 min.), and the bromine wholly expelled. Dilute to about 100 ml., add 2 ml. of concentrated nitric acid, and again boil. On adding silver nitrate to the hot solution the characteristic curdy *white* precipitate of AgCl is at once thrown down in quantity. The precipitate will be found to be immediately and completely soluble in ammonia (cf. AgBr).

It is of interest to examine the products of the above exps. (i) and (ii) under the microscope, and compare their appearance with that of a specimen of sodium bromate crystallized from water alone. The latter will consist almost entirely of well-formed tetrahedra, while the presence of the other salts induces the production almost exclusively of cubes, for the most part faceted at four of the eight corners.

Another interesting group of (homochemical) substances to study is formed by the three halide salts, KBr, $(NH_4)Cl$ and NaCl. Within this set the K and NH_4 salts are sufficiently nearly homoplastic to form a complete series of mixed crystals in all proportions of the constituent salts; on the other hand, neither of these salts mixes appreciably with NaCl in the solid state (cf. table 5).

In addition to works mentioned in the text, the following are recommended for reading and consultation:

James, *X-Ray Crystallography*. (Methuen's Physical Monographs, 1961.) A compressed summary, mainly of modern methods.
Bunn, *Chemical Crystallography*. (Clarendon Press, 1961.)
Wells, *Structural Inorganic Chemistry*. (Clarendon Press, 1950.)
Evans, *Crystal Chemistry*. (Camb. Univ. Press, in the press, 1962.)
Phillips, *Introduction to Crystallography*. (Longman, 1956.) A summary of classical crystallography.

Chapter III

SOLUTIONS AND SOLUBILITY

Introduction. It is convenient to treat this subject under three heads, distinguished by the nature of the system of which the solution forms a part: (1) *Simple systems*, in which one solution is in equilibrium with one or more pure phases (e.g. gases and water, salts and water, at normal temperatures). These are included in Part 1 of the present chapter. (2) *Mutual systems*, where two or more saturated solutions are in mutual equilibrium, and pure phases may be absent (e.g. partially miscible liquids, and equilibria between liquid and solid solutions) form the subject of Part 2. Completely miscible liquids are regarded as extreme examples of mutual solubility. (3) *Dilute solutions*, to which Chapter IV is devoted.

Although, as the introductory notes will show, the phenomena of solubility can be profitably discussed from first principles, without the aid of the phase rule, students should acquaint themselves with the bearing of this generalization on the examples studied during the course of their work. For this purpose the following may be mentioned:

Findlay, *The Phase Rule and its Applications*, revised and enlarged by Campbell and Smith (Dover Publications, 1951).

Hildebrand and Scott, *Solubility of Non-electrolytes* (Monograph Amer. Chem. Soc. No. 17, 1950), may also be recommended as a stimulating review of the whole subject, treated on the basis of Raoult's law.

(See also Chapter III, Part 2, p. 86.)

PART 1. SIMPLE SOLUTIONS

(1) The solubility of gases

A gas in equilibrium with its solution in a liquid with which it does not chemically combine presents the physically simplest example of a system containing solutions, although the concentration of the solution is sensibly dependent on *both* temperature and pressure (cf. Solubility of Salts, pp. 65 sqq.).

It must be assumed that the equilibrium is dynamic, i.e. that molecules of gas are continually exchanging places in the gas-space and the solution, and that the rate of *introduction* of the gas into the liquid is exactly equal to its rate of *evaporation* from the solution. The former rate is proportional simply to the number of molecules striking unit area of the liquid surface per second, and this in turn, from the kinetic theory, to the pressure. The rate of the opposing process (evaporation) will depend upon two factors, first the concentration, and secondly, the amount of energy that a dissolved molecule must possess in order to escape from the attractive forces of the surrounding molecules; this energy is equal to that liberated on introduction. If this energy does not itself depend on the concentration, we may write

rate of evaporation $= k_1 C =$ rate of introduction $= k_2 p$,

where the k's are constants of proportionality. Hence

$$p/C = \text{constant.}$$

This expression represents Henry's law, which will be valid as long as the work required to extract a dissolved molecule from its surroundings remains independent of the concentration, which means, while the solution is *dilute,* and the dissolved molecules on the average so far separated that the surroundings of each consist entirely of solvent molecules.

Hence only gases sparingly soluble at ordinary pressures (e.g. oxygen, exp. *e*, p. 60) obey the law at such pressures, but all gases would conform in the range of (low) pressure corresponding to dilute solutions, in the absence of chemical changes, which, however, are usually associated with great solubility (e.g. ammonia, sulphur dioxide and hydrogen chloride); but even gases very soluble at ordinary temperatures and pressures obey the law at such pressures and at such higher temperatures, that gaseous solubility is small. This effect of increased temperature (exp. f, p. 63) is to be correlated with the fact that energy is always liberated (as heat) when gases dissolve. Gaseous solubility may also be greatly reduced by the presence of dissolved electrolytes (exp. b, p. 57).

(Those familiar with the Maxwell-Boltzmann law of energy distribution may prefer to derive the laws of gaseous solubility directly from the relation

$$p = (\text{constant})\, Ce^{-L/RT},$$

where C is the concentration of the solution, and L the energy of solution per gram-molecule of gas.)

The solubilities of gases may be expressed in various ways:

(a) The coefficient of solubility (Bunsen): this is the volume of gas reduced to N.T.P. dissolved by unit volume of solvent at a specified temperature and under a pressure of 1 atm. (but not necessarily at N.T.P.). Its usefulness lies in the fact that it is proportional to the molecular concentration of the solution.

(b) For gases conforming to Henry's law—the volume dissolved by unit volume of solvent at a given temperature: this volume of gas will be independent of the pressure over the range of validity of Henry's law.

(c) The weight in grams of gas dissolved by *unit volume* of solvent. This mode of expression, analogous to that commonly used for solids, is only useful for very soluble gases.

The type of method used to determine gaseous solubility depends on the magnitude of the solubility. Only when the volume of gas dissolved is comparable with the volume of solvent is it convenient to measure the gas volume directly.

(a) To determine the solubility of carbon dioxide in water at room temperature

The apparatus (fig. 13) consists of a gas burette (see fig. 3, p. 13) and a separating funnel with rubber stopper and capillary inlet tube; connexion between funnel and burette is made through two *short* pieces of stout-walled rubber tubing, each with screw-clip (S_1 and S_2), and an interposed short section of capillary glass tubing. The separating funnel should have a capacity of about 100 ml. and be preferably of cylindrical form, with a short stem and well-fitting tap.

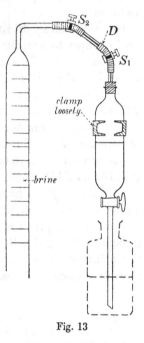

Fig. 13

(i) Determine the capacity of the funnel: fill to the rim with water, fit the stopper so that water issues through the capillary and no bubble remains below. Run out the water through the tap into a graduated cylinder.

(ii) Using brine (in which carbon dioxide is almost insoluble) as liquid in the burette, charge it to about the 200 ml. mark with pure carbon dioxide; the clip S_2 is then closed.

(iii) Prepare gas-free water in quantity more than sufficient to fill the funnel. This may be done by boiling out, but more quickly and effectively as follows: Place distilled water in a glass bottle so that the bottle is about two-thirds full; close with a rubber stopper carrying a tube and stout rubber connexion with screw-clip (if possible, choose a bottle in which the stopper and connexion of the funnel can be fitted for this purpose); connect to the water-pump and exhaust, at the same time vigorously shaking the bottle, until the contained water gives a pronounced 'hammer'. Close the clip and detach the bottle from the pump. The removal of dissolved gas is necessary, not because it would materially affect directly the solubility of carbon dioxide, but for the reason that during the experiment nearly all the dissolved gas (mainly air) would be liberated into the carbon dioxide above the water in the funnel, and the partial pressure of that gas considerably reduced.

(iv) Completely fill the funnel with gas-free water, as in (i), and connect to the burette; equalize levels if necessary and read the burette. Open the clip S_2 and run out through the tap about half the water in the funnel into a graduated cylinder, taking care to add brine to the burette to compensate for the gas withdrawn into the funnel. Note the volume of water run out. Shake the funnel as much as the connexions permit until the intake of gas slackens. Then raise the pressure slightly above atmospheric, close *both* clips and detach the funnel at D (fig. 13). Hold the funnel by the tap and neck to avoid heating the solution by the hand, and shake it vigorously for about 1 min. Connect again to the burette, open the clips, and adjust to atmospheric pressure; repeat the shaking after detachment as before. Proceed with alternate shaking and gas intake until the latter ceases. Finally, read the burette and the barometer, open the funnel and take the temperature of the solution.

Example

Capacity of funnel = 118 ml.
Water run from funnel = 60 ml. = gas-space in funnel.
Water remaining for solution = 58 ml.
Initial burette reading = 219·5 ml.⎫
Final burette reading = 105·0 ml.⎬ Difference = 114·5 ml.
Gas remaining in funnel = 60·0 ml.
Gas dissolved in 58 ml. of water = 54·5 ml.
Barometer height = 765 mm. Temperature of solution = 20°.
Aqueous tension at 20° = 17·4 mm.
Net pressure of gas in equilibrium with solution = 747·6 mm.
Solubility:

(a) Volume in unit volume of water = 0·940.

(b) Coefficient of solubility = $0·940 \times \dfrac{273}{293} = 0·876$.

(Henry's law assumed between 745 and 760 mm.)

Small errors are introduced into the above direct method by the slight solubility of the gas in brine, and in the rubber connexions. These may be avoided by estimating the amount of dissolved gas by titration, and confining the use of the gas burette to a test of saturation. The method of titration follows the lines of that in exp. *c*, p. 19, which should be read for details.

Place in a glass-stoppered bottle of about 200 ml. capacity 50 ml. of an approximately $N/5$ solution of sodium hydroxide. If necessary, attach an extension of capillary tube to the stem of the funnel in which the gas solution has been made, so that when the bottle is brought below the funnel the tube reaches to the bottom of the contained liquid. Discharge the gas solution into the alkali slowly and completely, but do not allow a bubble of the gas to pass into the absorbing solution.* Stopper the bottle and gently shake. Transfer with washings to a flask, add 20 ml. of reagent barium chloride solution and titrate with standard hydrochloric acid (about $N/4$), using phenolphthalein, as in the experiment quoted. Titrate a specimen of the original soda solution as a control after similar treatment with 20 ml. of barium chloride solution. The volume of carbon dioxide in the gas solution is then found from the difference in the titrations, assuming that 36·5 g. of hydrogen chloride are equivalent to 11,200 ml. of carbon dioxide (at N.T.P.).

(*b*) *Examine the effect of dissolved electrolytes on the solubility by using as solvents molar potassium chloride and semi-molar sodium sulphate.*

(*c*) *Use as solvent $N/10$ sodium carbonate, and find from the increased solubility the extent to which dissolved sodium carbonate is converted into bicarbonate by carbon dioxide at atmospheric pressure.*

(*d*) *To study the solubility at various pressures less than atmospheric, a satisfactory method is to dilute the carbon dioxide with hydrogen.*

* During this operation the gas-burette must remain connected to the funnel, with the clips open (see fig. 13).

A tap funnel of known capacity is again required, but the rubber stopper should carry two capillary inlet tubes, with rubber connexions and screw-clips, as shown in fig. 14. A mixture of hydrogen (or, less advantageously, air) with carbon dioxide in known volume proportions is prepared beforehand in a small aspirator with brine as working liquid. First fill the funnel to the rim with distilled water (in this experiment not necessarily air-free) and firmly insert the stopper and tubes. Connect tube t_1 to the aspirator and expel about half the water in the funnel through t_2 into a graduated cylinder by which

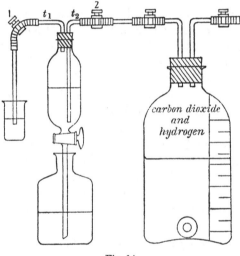

Fig. 14

the volume expelled is measured, and the volume remaining in the funnel so determined. Screw up clip 2 and well shake the funnel and contents: then screw up clip 1 and change the aspirator connexion to t_2. Insert into t_1 a tube dipping just below water contained in a beaker. Pass an excess of the gas mixture from the aspirator through the funnel, to replace completely that already above the water: close the clips, detach the funnel, and again shake vigorously. The procedure of renewing the gas mixture above the water and shaking is continued until on a final connexion to the aspirator (now with liquid levels carefully equalized), clip 1 being still closed, no gas passes into the funnel through t_2. The now saturated solution is run out through the tap of the funnel, t_2 being open to the aspirator and gas mixture at atmospheric pressure being allowed to fill the funnel, into 50 ml. of an $N/5$ solution of sodium hydroxide in the manner described on p. 57. The dissolved carbon dioxide is estimated by the titration method also described on p. 57. The barometer height and the laboratory temperature during the experiment should be recorded.

Example

The prepared mixture (2 l.) contained equal volumes of hydrogen and carbon dioxide.

13·6 ml. of $N/5$ hydrochloric acid were found to be equivalent to the carbon dioxide dissolved in 70 ml. of solution.

Barometer height $= 771$ mm. Temperature $= 19°$. Aqueous tension $= 16·3$ mm.

1 ml. of $N/5$ hydrochloric acid $= 2·24$ ml. carbon dioxide (N.T.P.).

Carbon dioxide dissolved at pressure 377 mm. in 70 ml.

$$= 13·6 \times 2·24 = 30·46 \text{ (N.T.P.)}.$$

Solubility (vol. of gas/vol. water) $= \dfrac{30·46}{70} \times \dfrac{760}{377} \times \dfrac{292}{273} = 0·938.$

(Cf. exp. at atm. pressure, p. 56; carbon dioxide obeys Henry's law fairly closely up to about atmospheric pressure.)

Table 7. *Solubility in water of carbon dioxide at 760 mm.*
(vol. of gas/vol. water)

$t°$	Solubility	$t°$	Solubility
15	1·074	21	0·920
16	1·042	22	0·896
17	1·016	23	0·872
18	0·989	24	0·850
19	0·965	25	0·828
20	0·942		

(e) *To determine the solubility of oxygen at room temperature*

The solubility of oxygen is found for water in equilibrium with (i) air, (ii) pure oxygen, each at atmospheric pressure. As oxygen is sparingly soluble and may be expected to follow Henry's law, the solubilities should be almost exactly in the ratio 1:5.

The volume dissolved in a limited amount of water, being too small to be estimated directly in an ordinary gas burette, is best found by the chemical method of Winkler.

When manganous hydroxide, prepared in situ from manganous chloride and sodium hydroxide, is thoroughly mixed with the solution of oxygen, the gas is completely extracted by oxidation of the hydroxide to (hydrated) sesquioxide, Mn_2O_3. On mixing the latter product with acidified potassium iodide the additional oxygen liberates its equivalent of iodine, which is then titrated with standard thiosulphate:

$$2MnO + O = Mn_2O_3$$
$$\text{white} \qquad \text{brown}$$

$$Mn_2O_3 + 6HI = 2MnI_2 + I_2 + 3H_2O,$$

$$11 \cdot 2 \text{ l. of oxygen at N.T.P.} = I_2 = 2Na_2S_2O_3$$

or

1 ml. $N/50$ thiosulphate $= 0 \cdot 112$ ml. dissolved oxygen, reduced to N.T.P.

Solutions required:

(1) 40 g. of $MnCl_2 . 4H_2O$ in 100 ml. of water.
(2) 30 g. of NaOH in 100 ml. of water.
(3) Standard sodium thiosulphate (about $N/50$).

(i) In addition to the apparatus of exp. *a*, p. 55, two glass-stoppered bottles, each of capacity about 200 ml., will be required. Prepare a supply of fully aerated water by drawing with the aid of the water-pump a current of air first through a tube containing soda-lime and then through distilled water.

Determine the capacity of each bottle as follows: Fill to the rim with water, allow the stopper to fall into place, then remove it and pour the water into a graduated cylinder. Take the temperature of the aerated water, and fill one of the bottles with it to the rim. Using a 1 ml. pipette with rubber tube as safety extension, discharge at the bottom of the bottle 1 ml. of the solution of sodium hydroxide (2) with as little mixing as possible; in the same manner discharge 2 ml. of the solution of manganous chloride (1). Allow the stopper to fall into place without including any bubble, and then make it secure by twisting. Mix the contents of the bottle by laying it flat on the bench and spinning it several times. Allow the brown precipitate to settle completely, and then open the bottle, and add first a few crystals of potassium iodide and then 5 ml. of concentrated hydrochloric acid (use a 5 ml. pipette, *with safety extension*), releasing at the bottom as before. Close the bottle and again mix by rotation. When all traces of brown precipitate have dissolved, transfer to a 400 ml. flask with washings, and titrate with $N/50$ thiosulphate.

(ii) Charge the gas burette with about 150 ml. of oxygen over water. Fill the funnel (of estimated capacity) with gas-free water, prepared as in exp. *a*, p. 55. Fit the stopper and close the clip. Prepare a second quantity of gas-free water, and with it fill one of the glass-stoppered bottles (capacity 200 ml.) to the rim and then set in the stopper, leaving no bubble. Connect the funnel to the burette, open both clips and draw in oxygen by running out water so as to leave 40 ml. in the funnel. Equalize levels in the burette, and then close both clips and detach the funnel. Shake as before to establish equilibrium, connecting only once again to the burette. Remove into a flask with a pipette 50 ml. of water from the bottle of gas-free water, and place the bottle below the funnel so that the stem of the latter dips to the bottom (extend the stem with glass tubing if necessary). Open the tap of the funnel and allow the contents to flow into the bottle, at the same time as oxygen at barometric pressure enters the funnel; discharge all the oxygen solution into the bottle, but do not allow a bubble of oxygen to pass through the water. Remove the bottle, 'top off' with gas-free water, and proceed to estimate the dissolved oxygen by Winkler's method as above. The precautions suggested are directed to preventing contact between the undiluted solution and air, to which the solution would rapidly lose oxygen; after the 5 times dilution there is no longer any danger of such loss.

Example

(i) *Water saturated with air.*

Capacity of bottle = 198·5 ml.

Temperature of saturated water = 16·0°.

$N/50$ thiosulphate required for iodine liberated = 11·75 ml.

Volume of oxygen (N.T.P.) dissolved in 198·5 ml. at 16°

$$= 0·112 \times 11·75 = 1·316 \text{ ml.}$$

$$\text{Solubility} = \frac{1·316}{198·5} = 0·0066.$$

(ii) *Water saturated with oxygen at barometric pressure.*

Capacity of funnel = 137 ml.

Water run out = 99 ml.

Water for solution = 38 ml.

$N/50$ thiosulphate required for iodine liberated = 11·5 ml.

Solubility calculated as above = 0·034.

Ratio of solubilities = 5·1 : 1.

Table 8. *The absorption coefficient of oxygen* (760 *mm.*)

Temp. °C.	Coefficient
10	0·038
11	0·037
12	0·036
13	0·0355
14	0·035
15	0·034
16	0·033
17	0·033
18	0·032
19	0·032
20	0·031

(*f*) *To determine the solubility of ammonia,* (i) *at* 0°, *and*
(ii) *at room temperature*

(i) A small bulb apparatus, of approximately the dimensions indicated in fig. 15, is blown from a length of quill
tubing, and weighed. Insert one tube of the bulb into a
boiling tube containing water or other liquid (e.g. alcohol) to
be used as solvent (fig. 15 *b*). By suction draw liquid up into
the bulb until the latter is about two-thirds full. Arrange the
apparatus in a bath of mixed ice and water, and connect the
wider end with rubber tube to a supply of ammonia (fig. 15 *a*).

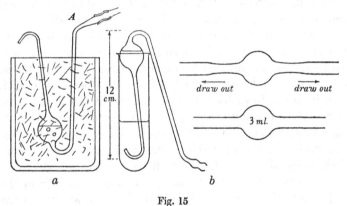

Fig. 15

(This is most easily obtained by *gently* warming 0·880 ammonia
solution in a flask; if solubility in alcohol is being determined,
the gas must be dried with a lime column.) Allow a stream of
gas to pass into the bulb, controlling the rate so that little
water is expelled from the exit tube. The bath should be
stirred frequently to prevent the accumulation of water at 4°
(maximum density) round the bulb in the lower part of the
bath. When water is used as solvent at least an hour's treatment with the gas is essential to reach saturation, but this
period can be shortened to half an hour if 0·880 ammonia
solution instead of water is placed in the bulb at the beginning.
The bulb should in this case be filled by suction through a
wash-bottle of water, or by a short application of the water-

pump, but *on no account* by direct suction with the mouth. When saturation is judged to be reached, disconnect the ammonia supply, but do not remove the bulb from the bath. With the aid of a spirit lamp draw off and seal at about the point *A* in fig. 15*a* (see method under exp. *a*, p. 20); retain the drawn-off portion for weighing. *Very* lightly fuse over the other, finer, end of the apparatus. Inspect the seals with a hand lens, and if they are found satisfactory, lift the bulb from the bath, allow to reach room temperature, wipe dry, and then weigh, together with the drawn-off portion. Note the height of the barometer.

Place in a porcelain dish 50 ml. of standard (2 N) sulphuric acid (use a safety extension on the pipette), and dilute with water so that when the sealed bulb is placed in the dish it can be wholly submerged. Hold the bulb under the acid,* and break one seal by pressure between the jaws of crucible tongs. When most of the ammonia has been absorbed in the acid, fracture the bulb with a pestle to ensure complete absorption. Titrate the liquid with sodium hydroxide solution of arbitrary concentration (about normal), using methyl *red* (methyl orange if methyl red is not available). Titrate separately 20 or 25 ml. of the original acid with the alkali and the same indicator.

(ii) Repeat the determination, omitting the ice from the bath. Take the final temperature of the bath as that of the solution. The bath should be frequently stirred.

Example

Weight of the bulb = 1·9972 g.
Weight of the bulb with solution = 3·5162 g.
Weight of the solution = 1·519 g.
Temperature = 0°. Barometer height = 770 mm.
20 ml. of sulphuric acid, 109·7 g. H_2SO_4 per litre, required 44·0 ml. of the solution of sodium hydroxide for neutralization.
50 ml. of the same acid, after partial neutralization with the ammonia from the prepared solution, required 69·3 ml. of the alkali solution for neutralization.

Sulphuric acid equivalent to ammonia $= 50 - \left(69\cdot3 \times \dfrac{20}{44}\right) = 18\cdot5$ ml.

* It is advantageous to cool the bulb again in ice before breaking the seal under the acid.

Weight of ammonia contained in 1·519 g. of solution

$$= 18 \cdot 5 \times 0 \cdot 1097 \times \frac{17}{49} = 0 \cdot 704 \text{ g.}$$

Water in solution, by difference $= 0 \cdot 815$ g.
Solubility (g. NH_3 per 100 g. water) $= 86 \cdot 4$.

Table 9. *Solubility of ammonia at 760 mm. in water*

(g. NH_3 per 100 g. of water)

Temp. °C.	Solubility
0	87·5
...	...
12	64·5
14	61·2
16	58·2
18	55·4
20	52·6

(iii) The solubility of ammonia in methyl or ethyl alcohol may be determined by the same method. Under the same conditions, the solubility in methyl alcohol, expressed as weight of gas in 100 g. of solvent, is somewhat greater than in water; in ethyl alcohol, about one-fifth of that in water. If the solubilities are compared on the more rational basis of gram-molecules of solvent, with water = 18, then the solubility in methyl alcohol appears as double, and that in ethyl alcohol as about one-half of the solubility in water.

(2) The solubility of salts in water

(*Note on the sign of energy changes.* By general agreement the sign to be attributed to an energy change in a physico-chemical system is decided by considering the direction of the change undergone by the *system* when the conditions of the equilibrium are maintained. In the discussion to follow constancy of temperature is the condition implied, since solubility depends upon temperature. To secure such constancy, some sort of thermostatic control must be assumed, which must remove heat liberated, and supply heat absorbed, in the system. In the former event the system loses heat, and the change is reckoned *negative*; in the latter, *positive*, for the system gains heat. In short, if in the absence of effective temperature control the temperature *rises*, the heat change is *negative*, and conversely.)

When gases dissolve, the energy change, always negative, is solely due to the interaction of the gas and solvent molecules (see Solubility of Gases, p. 54). In discussing the equilibrium between *solids* and their solutions in liquids, we must assume that a large energy input (i.e. positive energy change) will in general be required to extract a molecule of the solid from the pure solid phase. It may be calculated that the removal of an ion (e.g. Na^+ or Cl^-) from a crystal of the corresponding electrolyte (sodium chloride), wherein the ions are closely surrounded by oppositely charged ions, requires the expenditure of energy of the order of 100,000 cal./g.-ion. Unless such a consumption of energy can be to a large extent compensated by a liberation when the ions enter the solvent, appreciable dissolution cannot take place.

Water exhibits prominently the characteristics of a class of compounds called *polar*, because their molecules behave as though they had permanent electric charges (not to be confused with ionic charge) on opposite sides. This feature, which may be symbolized in water by the structural formula $^+H\diagup\overset{O^-}{}\diagdown H^+$, is responsible for the 'abnormal' properties of many liquids (see, for example, the notes on constant-boiling mixtures, p. 83). Kations, on entering water, at once become enveloped in water molecules oriented with oxygen towards the ion, and anions become surrounded with molecules oriented in the opposite sense. Both ions are thus brought into an environment similar to that they have left in the crystal, and energy is liberated (L_1) comparable in amount with that demanded for escape from the crystal (L_2).

From this analysis, we may draw conclusions in harmony with observed facts: (1) electrolytes only dissolve appreciably in polar solvents, particularly in water, (2) the heat change,

$\Delta H_{\text{sol.}}$, which is only the balance of much larger energy changes $(\Delta H = L_2 - L_1)$, may be positive or negative, which leads by the principle of Le Chatelier to (3) the solubility of electrolytes may increase or decrease with rise of temperature, but since we may usually expect L_2 to be greater than L_1, the solubility will usually rise with rise of temperature. Further, it may be predicted that salts already much hydrated in the crystal (e.g. $Na_2CO_3 . 10H_2O$, see exp. b, p. 72) will pass into solution with relatively large positive heat change.

The laws of energy distribution enable us to deduce the change of solubility with temperature as follows:

We assume, in analogy with gaseous solutions (p. 53), that in a saturated solution of a solid the rate of deposition upon the solid is exactly equalled by the rate of dissolution from the solid:

Rate of deposition $= k_1$ (number of dissolved molecules with energy L_1 to escape from the solvent environment)
$$= k_1 C_s e^{-L_1/RT}.$$

Rate of dissolution $= k_2$ (number of molecules with energy L_2 to escape from the lattice attraction)
$$= k_2 C_0 e^{-L_2/RT}.$$

C_s and C_0 are the concentrations in solution and solid respectively: C_0 may be taken as independent of temperature.

Therefore $\quad C_s = (k_2/k_1) C_0 e^{(L_1-L_2)/RT}$
$$= \text{constant } e^{-\Delta H/RT},$$
$$\frac{d \ln C_s}{dT} = \frac{\Delta H_{\text{sol.}}}{RT^2},$$

or, alternatively, $\quad \ln C_s = A - \Delta H_{\text{sol.}}/RT,$

where A is a constant.

The solubility of a solid is commonly stated as the weight dissolved in 100 g. of *solvent*. From the solubility thus expressed the molar fraction can be computed, but the molecular concentration cannot be found without a knowledge of the specific gravity of the solution. Solubilities of hydrated salts are stated as weight of *anhydrous* salt dissolved by 100 g. of water, and if it is desired to calculate the solubility in terms of the hydrated salt, the hydrate water must be deducted from the 100 g.

(*a*) *To determine the solubility of ammonium oxalate
in water*

(i) *The solubility at* 0°.

With a pestle and small mortar grind up about 3 g. of the salt with 25–30 ml. of water, added in portions; the salt should remain in excess. After thus obtaining a solution nearly saturated at room temperature, pour the contents of the mortar into a boiling tube, which is then immersed during 30 min. in a bath of ice and water. Shake or stir the solid with the liquid in the tube frequently, and also stir the bath. Then, without taking the tube from the bath, insert into it a fairly tight plug of clean cotton-wool ¾ to 1 in. deep. By applying firm and steady pressure with a glass rod to the centre of the plug pass the latter down the tube and through the liquid, which is thus filtered free from solid without appreciable change of temperature (fig. 16). With-

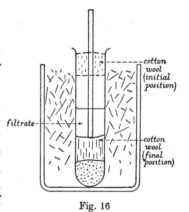

Fig. 16

draw 10 ml. of the solution by a pipette, deliver into a flask, add about 30 ml. of dilute sulphuric acid, and titrate at 80° with $N/10$ potassium permanganate to a faint permanent pink:

$$5(NH_4)_2C_2O_4 = 2KMnO_4$$
$$620 \qquad\qquad 316$$

As the solubility at 0° amounts to only about 2 g./100 g., the weight of the sample can be taken as 10 g. in calculating the solubility.

(ii) *The solubility at* 100°.

Place in a boiling tube 11 g. of the finely powdered salt, and 25 ml. of water. Insert a glass ring stirrer, and close the tube with a grooved cork carrying a thermometer. Immerse the tube as deeply as possible in hot water contained in a beaker. Stir the mixture in the tube vigorously as the water-bath is raised to boiling-point.

Place the stoppers on a Landolt pipette (fig. 17 b) and weigh it. When the solution has been stirred at intervals for at least 10 min. after reaching 100°, proceed to extract a sample with the aid of the pipette in the following manner (during these operations the bath should be continuously boiling). Remove the stoppers from the pipette, and the cork and thermometer from the solution tube, which is temporarily closed with cotton-wool. After extracting the thermometer, insert the upper stem of the pipette through the bore in the cork and replace the upper stopper of the pipette, which is then rather slowly introduced into the solution tube, and set in position as shown in fig. 17 a.

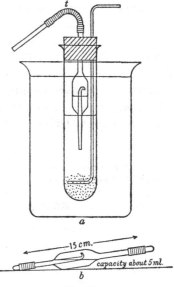

Fig. 17

Air expanding out of the pipette will prevent solution from entering the lower stem and choking it with deposited salt before it has taken the temperature of the solution. After a short time the upper stopper can safely be replaced by an open rubber tube *t* (fig. 17 a). The solution is again stirred for about 5 min., and liquid expelled once or twice from the lower stem by *gently* blowing through the rubber tube. Allow the solid to settle fully, and then by *cautious* suction draw into the now heated pipette about 2 ml. of the solution (the rubber connexion should be long enough for the intake of solution to be clearly watched).

The cork and pipette can now be lifted from the tube; much time will be saved later if, during the removal, the lower stem is kept clear of solution by gentle blowing through the rubber tube. Rotate the pipette in a horizontal position as it cools, to cause the solid to deposit round the sides. Wipe the outside free from solid, replace both stoppers, and when quite cold, weigh.

Immerse the pipette, *upper stem downwards*, in about 40 ml. of warm dilute sulphuric acid contained in a boiling tube, when the solid should rapidly dissolve, especially if the pipette is moved up and down. Transfer the liquid to a 200 or 250 ml. graduated flask, and unite all washings with the main solution, which is made up with distilled water. Take 50 ml. of the solution by ordinary pipette, add about an equal volume of dilute sulphuric acid, and titrate at 80° with $N/10$ potassium permanganate as before in exp. (i).

Example

(Determination at 100°.)

Weight of sample $= 3 \cdot 071$ g.

The sample was made up to 200 ml., and 50 ml. taken for titration required 33·7 ml. of $KMnO_4$ solution (2·79 g./l.).

Hence 3·071 g. of solution contains

$$\frac{2 \cdot 79}{1000} \times \frac{620}{316} \times 33 \cdot 7 \times 4 = 0 \cdot 744 \text{ g. salt.}$$

Water in solution $= 2 \cdot 327$ g.   Solubility $= 100 \times \dfrac{0 \cdot 744}{2 \cdot 327} = 32 \cdot 0$ g.

Percentage of salt in solution $= 24 \cdot 2$.

(iii) *The solubility curve.*

Use the formula $\log_{10} S = A - B/T$ (see p. 67), where S is the *percentage* of salt in the solution and A and B are constants, independent of temperature, to construct a solubility curve in the following way. Set out as abscissae on squared paper the values of $1/T$ given below, and mark the corresponding values of t against the points. Take as ordinates $\log_{10} S$. Draw a straight line through the values of $\log_{10} S$

determined at $0°$ ($T = 273°$), and at $100°$ ($T = 373°$). Read off $\log_{10} S$ and thus evaluate S, at intermediate temperatures (fig. 18).

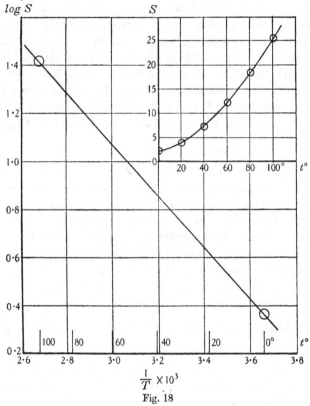

Fig. 18

Table 10. *Ammonium oxalate*

$t°$	$T°$	$1/T$	Solubility (see p. 72)
100	373	0·00268	34·6
80	353	0·00283	22·4
60	333	0·00300	14·0
40	313	0·00319	8·2
20	293	0·00341	4·45
0	273	0·00366	2·36

Compare the results with the data given in table 10, in which the solubilities (salt/100 g. solvent) have all been determined *experimentally* at the temperatures shown.*

In the formula given above $B = \Delta H_{sol.}/2 \cdot 3R$, where $\Delta H_{sol.}$ is the heat of solution (assumed independent of temperature) per gram-molecule of solute dissolving in the nearly saturated solution (cf. p. 67), and R is the constant of the gas law. Determine the slope of the plotted straight line, assume $R = 2$ cal., and so evaluate $\Delta H_{sol.}$. Note that since the formula contains log S and not S itself, the units in which S is expressed do not affect the calculated value of $\Delta H_{sol.}$.

(b) The system sodium carbonate-water

Many hydrated salts suffer a partial or complete loss of their water of crystallization on heating to temperatures well below 100°, even when they are in contact with their aqueous solutions. The liberation of water causes a partial liquefaction on heating the originally dry hydrate. As explained in the notes above, highly hydrated salts dissolve with marked absorption of heat; a diminution of the hydration will inevitably lessen this heat absorption, and may even cause a change of sign to heat liberation (as in the case of $Na_2CO_3 . 10H_2O$ and $Na_2CO_3 . H_2O$ below). Thus from the relation $\dfrac{d\ln C_s}{dT} = \dfrac{\Delta H}{RT^2}$ (p. 67) the slopes of the solubility curves for different hydrates will decrease with decreased hydrate water, and we shall expect a different curve for every hydrate stable in contact with solution.

These different curves must intersect on the general solubility diagram (fig. 19), and an intersection define a certain temperature and solubility, *uniquely* associated with the presence of *two* hydrates simultaneously in equilibrium with the same solution, and therefore with each other. If then any conditions arise which allow the two hydrates to be both present in contact with solution, this particular temperature and composition of solution must be spontaneously established.

* Hill and Distler, *J. Amer. Chem. Soc.* **57**, 2203 (1935).

At temperatures below that of an intersection only one hydrate is in stable equilibrium with solution, and above the intersection temperature, another, lower hydrate. Therefore

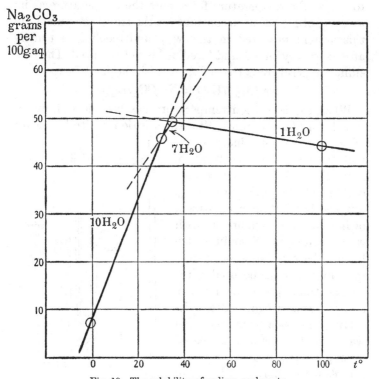

Fig. 19. The solubility of sodium carbonate.

this intersection marks a change from one hydrate to another, and is accordingly termed a *transition point*. The temperatures of these points are so well defined that some of them are used as fixed points in thermometry (p. 6).

(i) *To determine the transition temperatures of the system sodium carbonate-water, and the solubilities at these temperatures.*

Introduce 60 g. of *freshly* powdered $Na_2CO_3 . 10H_2O$ in portions into a boiling tube immersed in a water-bath kept at about 50°. When the contents of the tube have liquefied sufficiently stir regularly and keep at about 50° for 15 min. Then raise the tube out of the bath, and arrange a thermometer with its bulb dipping to the centre of the mixture. Continue to stir as the temperature falls; note the temperature at intervals until it becomes constant (at the transition point). In this system prepared in this way undercooling is usually absent or very slight, and 'seeding' is not required. The constant temperature is that of the transition point

$$Na_2CO_3.7H_2O \rightleftharpoons Na_2CO_3.H_2O.$$

When the transition temperature has become well established (for about 5 min.), filter the solution in situ by means of a cotton-wool plug, as explained in exp. a (i) (firm pressure is required). Withdraw a sample of the solution in a previously weighed Landolt pipette,* which, owing to the readiness with which solutions of sodium carbonate become supersaturated, may in this case be used without heating. Wipe the pipette, stopper and weigh. After dissolving out the contents with water, in the manner indicated in exp. a (ii), titrate the whole sample with a normal solution of hydrochloric acid (use methylorange).

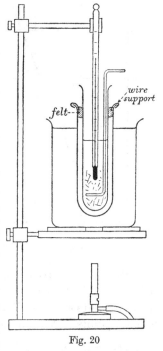

The transition point

$$Na_2CO_3.10H_2O \rightleftharpoons Na_2CO_3.7H_2O$$

is best observed as a thermal arrest on *heating* the decahydrate

Fig. 20

in contact with solution. Weigh out 40 g. of the powdered decahydrate (to the nearest gram). Place about half of the salt in

* See exp. a (ii), p. 69, for detailed instructions on the use of the Landolt pipette.

a boiling tube immersed in a water-bath kept at nearly 30°
but not above. Add 4 ml. of water and well stir with the salt;
as the temperature of the mixture rises and it becomes increasingly more fluid, stir in the remainder of the salt. When
the temperature has risen to nearly 30°, support the tube
inside a larger one also immersed in the water-bath, and continue to stir regularly. The outer tube serves as an air-jacket to
ensure slow heating of the inner (fig. 20). Raise the temperature of the bath, ultimately to about 50°, and note at
intervals the temperature of the mixture on a thermometer
dipping into it as before. When the constant transition temperature has become well established (at about 32°), withdraw
the inner tube somewhat from the air-jacket, filter in situ,
take a sample of the solution, and analyse, as before.

(ii) *The solubility of sodium carbonate at* 100°.

For this determination the detailed instructions given in
exp. *a* (ii) should be closely followed, except that 60 g. of the
decahydrate are heated to 100° without addition of water.
The solid remaining in excess at 100° is the monohydrate.
Estimate the sodium carbonate (Na_2CO_3) in the sample by
titration with a normal solution of hydrochloric acid as above.

(iii) *The solubility of sodium carbonate at* 0°.

Dissolve 20 g. of the decahydrate in 60 ml. of water. Fill a
boiling tube to about half its capacity with the resulting
solution, and cool the whole to 0° by immersing in an ice-water
bath. Stir vigorously, and if crystals do not separate, relieve
the supersaturation by the addition of a small quantity of the
solid decahydrate. Keep the mixture at 0° for 15 min., confirming the temperature on a thermometer dipping into the
tube, and occasionally stirring both the mixture and the bath.
Then filter in situ with cotton-wool, sample with a Landolt
pipette, weigh and analyse by titration as before.

Plot the results of the above experiments on squared paper,
taking solubility as ordinates and temperature as abscissae.
Construct an approximate solubility diagram for the system
by joining the points with straight lines (fig. 19). The slopes

of the lines should vary in agreement with the predictions in the introductory notes. It will also be noted that the stable existence of the heptahydrate is limited to the temperature range 32·1–35·3°. The steepness of the solubility line for the decahydrate is associated necessarily with a large heat absorption on dissolution. For the direct calorimetric determination of this heat of solution see exp. *b*, p. 172.*

Example

Transition temperature of $Na_2CO_3.7H_2O \rightleftharpoons Na_2CO_3.H_2O$.

Temperature	
1 min. intervals	2 min. intervals
41·0°	37·5°
40·0	36·1
39·1	35·0 (Slight undercooling)
38·7	35·3
	35·3 Transition
	35·3 temperature

Solubility at 35·3°.

A sample of the filtered solution weighing 5·796 g. required 36·2 ml. of *N* HCl for neutralization. It therefore contained 1·92 g. Na_2CO_3 and, by difference, 3·876 g. water.

Solubility = 49·5 g.†

Solubility at 100°.

The sample weighed 4·980 g. and required 29·0 ml. of *N* HCl for neutralization.

Solubility (calculated as before) = 44·8 g.†

Solubility at 0°.

By similar methods this was found to be 7·2 g.†

* A more accurate solubility diagram may be drawn by using the method (iii), p. 70, or by incorporating the data on p. 44.

† Kobe and Sheehy, *Ind. Eng. Chem.* **40**, 99 (1948) give: 0°, 7·00 g.; 35·3°, 49·5 g.; 100°, 44·7 g.

SOLUTIONS AND SOLUBILITY (*cont.*)

PART 2. MUTUAL SOLUBILITY

As long as experimental studies were largely confined to solutions in which a clear and logical distinction could be drawn between *solvent* and dissolved substance or *solute* (e.g. the simple solutions of Part 1), it remained possible to regard the solvent as merely providing the space in which the solute could behave as a gas. In such a scheme the (volume) concentration $C = n/V$, in which n is the number of gram-molecules dissolved in volume V, was the appropriate mode in which to express the composition of the solution.

The extension of experimental work to examples of mutual solubility enforced the abandonment in principle of all distinction between solvent and solute, and demanded an equal treatment of each constituent of the solution, for which the appropriate expression of composition is by *molar fractions*, $N_1 = n_1/(n_1 + n_2 + n_3 + \text{etc.})$, and similar ratios N_2, etc., for the other constituents.

The fact that molar fractions when small become approximately proportional to the concentration still leads to confusion, as when attempts are made to conjoin the consideration of lowering of vapour pressure on a basis of molar fraction, as originally laid down by Raoult (Chapter IV, p. 113), with a treatment of the closely related osmotic pressure on a basis of concentration ($\Pi = RTC$), although it has long been known that even for fairly dilute solutions this relation only very roughly reproduces the experimental measurements.

It can be asserted that in general the relations between the properties of solutions and their composition take their simplest and most significant form when the latter is expressed in molar fractions, although in some cases, where Raoult's law is inapplicable (e.g. solubility of gases, partition phenomena), it may still be necessary to reason in terms of concentration.

In calculating molar fractions (for binary mixtures), proceed as follows: (1) tabulate w_1/w_2, the ratio by weight; (2) multiply the values of w_1/w_2 by the (constant) factor M_2/M_1, to obtain the molar ratio r; (3) use a table of reciprocals to obtain $N_2 = 1/(1+r)$. Note that $N_1 + N_2 = 1$.

(1) Experiments on distillation

In discussing mixtures of liquids it is generally convenient to consider the properties as the sum of partial properties, attributed to the constituents separately; for example, the total vapour pressure exerted by the mixed vapours over a mixture of benzene and acetone is regarded as the sum of partial pressures of benzene and acetone; the total volume as the sum of partial volumes, and so on:

$$p_A + p_B = p, \quad v_A + v_B = v. \tag{1}$$

Raoult's law in its general form gives the ideal relation between partial properties and the composition; thus for partial pressures we have

$$p_A = N_A \cdot P_A, \quad p_B = N_B \cdot P_B, \tag{2}$$

where P stands for the vapour pressure over the pure constituent, and N is the molar fraction of the constituent (for explanation of molar fraction, see p. 77). The law may be expressed in words by saying that each constituent (of a binary mixture) acts as a simple diluent to the other. Fig. 21 a

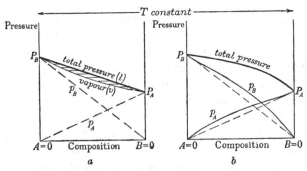

Fig. 21. a, ideal; b, showing positive deviation.

shows a graphical representation. In regard to most actual liquid mixtures, Raoult's law is applicable with exactitude to constituent 1 only when constituent 2 is present in dilute solution, i.e. N_2 is small, and conversely; but in general the law provides the only rational basis on which to interpret the behaviour of liquid mixtures. Deviations from the law are reckoned *positive* or *negative*, according as the actual partial vapour pressure is greater or less than that given by the ideal (linear) relation (2); the majority of liquid mixtures exhibit positive deviation (fig. 21 b).

If the ideal law be assumed to hold for all values of the molar fraction, the total pressure $p = p_A + p_B$ will be represented in the diagram (fig. 21 a) by a straight line l joining the pressures of the pure constituents, but the composition of the vapour exerting the total pressure will be given, even for an ideal mixture, by a curve (v) lying below the pressure-composition line l. The general truth of this may be deduced from the law as follows:

From (2)

$$p_B/p_A = (N_B/N_A)_{\text{liq.}}\, P_B/P_A\,;$$

by the kinetic theory

$$p_B/p_A = (n_B/n_A)_{\text{vap.}} = (N_B/N_A)_{\text{vap.}},$$

where n stands for number of gram-molecules. It follows that if, as in fig. 21 a, P_B is greater than P_A, i.e. B is more volatile than A, then $(N_B/N_A)_{\text{vap.}}$ is greater than $(N_B/N_A)_{\text{liq.}}$.

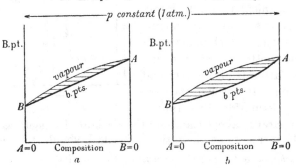

Fig. 22. *a,* ideal; *b,* corresponding to *b* fig. 21.

This result may be put into words—*the vapour is always richer in the more volatile constituent*—and on it depend all processes for the separation of liquid mixtures by distillation. When the result is expressed in a boiling-point diagram (fig. 22 a, b), the curve of vapour composition will lie above the boiling-point composition line. If, in a system yielding a constant-boiling mixture (minimum boiling-point), we regard that particular mixture as itself the most volatile constituent, the above rule is applicable also to such systems (see further, exp. b, p. 84).

(a) To determine by distillation the boiling-point diagram for mixtures of benzene and acetone

The ordinary apparatus for distillation is required: Wurtz flask, water-cooled condenser, and thermometer ($\frac{1}{2}°$ or $1°$ graduation). Six dry test-tubes, numbered in order with labels, and supported in a rack, are needed for receiving the distillate. Unless elaborate apparatus is used,* a certain amount of back condensation and consequent fractionation is unavoidable, but it may be reduced by choosing a flask with short neck and low side-tube, and further by enclosing it in a circular shield (fig. 23), improvised from a canister by cutting holes about an inch in diameter in the bottom and the lid, which latter is also cut in half to fit round the neck of the flask during distillation. In setting up the apparatus avoid clamping the flask at the neck, as this greatly increases condensation. The flask should be directly

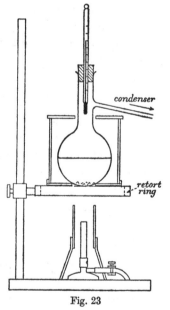

condenser

retort ring

Fig. 23

* *Technique of Organic Chemistry*, vol. IV. *Distillation*, Interscience Publishers, 1951, p. 8.

heated by a *small* flame, and not by a water-bath, as it is impossible with this to regulate the heating with nicety.

Place in the flask a mixture of 60 ml. of benzene (b.p. 80°) and 40 ml. of acetone (b.p. 56·2°). Add fragments of porous earthenware to facilitate smooth boiling, and set the thermometer with the bulb just below the level of the side-tube. Commence heating the flask without the lid of the shield in position, so that the contents of the flask can be observed. As the liquid comes near to the boiling temperature regulate the flame so that vapour just fails to enter the side-tube, and then set the lid of the shield in position. The decreased loss of heat will now generally suffice to start the distillation without increasing the flame, which may indeed need diminishing. Note the temperature registered when the first drop falls from the condenser. Distil steadily at a rate of about 1 drop issuing from the condenser per second. Collect the distillate in the test-tubes in turn, changing the receiver by moving the test-tube rack forwards at the temperatures shown in the example below. As the distillation proceeds, carefully mark the final levels of the fractions in the test-tubes with adhesive label (see below for measurement of the volumes of distillate). Remove the flame when the temperature reaches the boiling-point of benzene (80°). The whole distillation should occupy 20–30 min.

Analysis of the fractions. Transfer fraction 1 as completely as possible to a burette already containing 20–30 ml. of water. Cork the burette, and by rocking it gently cause the fraction to flow up and down the tube several times; violent shaking is to be avoided, as the very small droplets of benzene thus formed separate only very slowly (addition of dilute hydrochloric acid somewhat hastens separation). After the benzene mixture has all floated up loosen the cork and run away the aqueous solution of acetone from below. Wash in the same way with at least two more portions of water, after which treatment the liquid remaining may be taken as benzene free from acetone, and its volume read (as the lower meniscus is very curved read to its lowest point, and deduct 0·1 ml. from the reading). Repeat the method of analysis with the other fractions.

Finally ascertain the total volume of each fraction by running water from the burette into each test-tube receiver up to the paper mark. As there is no volume change on mixing benzene with acetone, the volume of the latter may be obtained by difference.

The composition of the liquid remaining in the flask is found by deducting the respective amounts of benzene and acetone distilled over; for example, the mixture at the beginning contained 60 ml. of benzene and 40 ml. of acetone; 6·6 ml. of benzene and 11·0 ml. of acetone (composition of fraction no. 1) had distilled when the temperature reached 64°: hence the composition of the liquid boiling at this temperature is (60 − 6·6) = 53·4 ml. of benzene, to (40 − 11·0) = 29 ml. of acetone.

Tabulate the results, and plot a curve connecting liquid

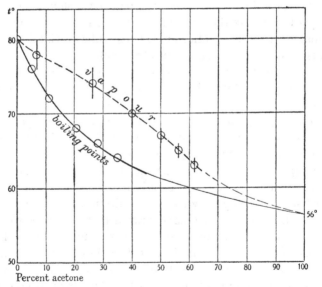

Fig. 24. *Benzene and acetone.* [Temperature ranges of fractions shown by vertical lines; approximate vapour line by drawing a smooth curve through mid-points of these lines.]

composition (expressed in volume percentage) with boiling-point (fig. 24). The vapour composition cannot be set similarly on a definite curve as we have by the above working only the average composition for a range of temperature, but an approximate attempt can be made, as shown.

Example

Original mixture = 60 ml. benzene (B) + 40 ml. acetone (A)

Temp. range °C.	Distillate (ml.)				Liquid in flask (ml.)			
	B	A	Total	Vol. % acetone	B	A	Total	Vol. % acetone
62–64	6·6	11·0	17·6	62	53·4	29·0	82·4	35
64–66	9·4	12·0	21·4	56	44·0	17·0	61·0	28
66–68	7·4	7·5	14·9	50	36·6	9·5	46·1	20·5
68–72	8·9	5·9	14·8	40	27·7	3·6	31·3	11·5
72–76	7·3	2·6	9·9	26	20·4	1·0	21·4	(5·0)
76–80	16·0	1·1	17·1	6·5	—	—	—	—
Total	55·6	40·1	95·7					

Benzene (almost free from acetone) left in flask + losses in manipulation = 4·3 ml.

(b) Constant-boiling mixtures

Suppose (1) that to a quantity of nitrogen 'peroxide' $(NO_2 + N_2O_4)$, contained in a small volume at atmospheric pressure, air is added, the total pressure being kept atmospheric as the total volume increases. The 'peroxide' will undergo 'dilution', but also dissociation; the partial pressure of the peroxide will consequently decrease less rapidly than if no lighter molecules could be produced. Consider (2) the effect of mixing the peroxide with a comparable volume of another dissociable gas: each will now undergo dilution and dissociation, and the partial pressure of each be greater than it would otherwise be.

The foregoing cases show much analogy with what takes

place when (1) benzene or other similar liquids, (2) water, are mixed with alcohols. The abnormally high boiling-points of most liquid hydroxy-compounds, and many other 'abnormal' properties, suggest that the effective molecular weight is much greater than would correspond to the simplest formula, and some form of association is presumed.* On 'dilution' by addition of benzene, dis-association must occur, and the decrease

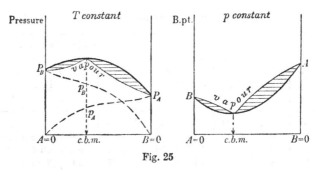

Fig. 25

of partial pressure due to dilution be offset by an increase of volatility, due to the increase of lighter molecules. Conversely, the strong attraction between the alcohol molecules exerts a sort of 'squeezing-out' effect on the diluent, whose volatility is thereby also increased. The lines of partial pressure thus exhibit unusually large positive deviation (fig. 25), and on summing to obtain the total pressure we may find (but not inevitably) a maximum in the total pressure curve.

Further consideration shows that the vapour composition line lies in the position shown in fig. 25, and therefore the mixture of maximum vapour pressure will vaporize without change of composition, i.e. it will boil at constant temperature like a pure substance. Inspection of a list† of the many pairs of liquids giving such constant-boiling mixtures of minimum boiling points reveals that at least one constituent is hydroxylic in the majority of cases. It will be clear from the figure that a maximum is most likely to be produced when the points P_A

* Ewell et al., Ind. Eng. Chem. **36**, 871 (1944).

† Horsley, ibid, Analytical Ed. **21**, 831 (1949).

and P_B are not at widely different levels, i.e. the boiling-points are not far apart, and that the position of the maximum, giving the composition of the constant-boiling mixture, will also depend on the relation of the boiling-points. The following data on mixtures of the lower alcohols and water illustrate these considerations:

Alcohol	Boiling-point	Constant-boiling mixture
Methyl	64°	None
Ethyl	78°	4·5 % of water
n-Propyl	92°	21 % of water

A convenient pair of liquids to study is benzene (b.p. 80°) and methyl alcohol (b.p. 64·7°) with constant-boiling mixture of b.p. 58·4°. Proof that the minimum-boiling mixture has constant composition is most easily obtained by distilling mix-tures of the liquids through a fractionating column, for which any ordinary type may be used, or one improvised as shown in fig. 26. The most effective filling material consists, not of the glass beads or short tubes commonly used, but of strips of thin aluminium foil, cut about 4 cm. long and 5 mm. wide, rolled into small spirals.

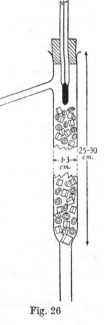

Two mixtures, in the proportions shown, should be distilled separately:

		Methyl	
Benzene		alcohol	
ml.		ml.	Final temp.
(1) 80		20	b.p. benzene
(2) 30		50	b.p. methyl alcohol

Receive the distillate in test-tubes as in (a), collecting about three fractions at the minimum temperature (58·4°, corrected),

Fig. 26

and one after the temperature has risen sharply to the final temperature. Analyse the fractions as in (a). The constant-boiling mixture should have 61·4 vol. % of benzene (see also exp. c, p. 24, Chapter I).

(2) The miscibility of liquids*

The following table, showing the temperature conditions necessary to ensure complete miscibility (in all proportions) of certain pairs of liquids, illustrates the principle that miscibility depends on likeness of molecular structure. The sign ∞ indicates complete miscibility throughout the range of existence of the two liquids:

	Aniline	Toluidine	
		Ortho-	Meta-
Benzene	∞	∞	∞
Cyclohexane	Above 31°	∞	∞
Dimethylcyclo-hexane	Above 49°	∞	∞
Paraffins	Above 70°	Above 11°	Above 9°

The more dissimilar the molecular structures of A and B, the greater is the (positive) deviation from Raoult's law exhibited by mixtures of A and B, but such deviations are always diminished by rise of temperature. Also, in all cases, Raoult's law applies to A (or B) when only small amounts of B (or A) are dissolved. Fig. 27 a, b, c summarize these considerations for the vapour pressures of the constituents at rising temperatures. At the lower temperature T_1 the deviation is of such magnitude that two mixtures (m_1 and m_2) have the same vapour pressure, and are consequently in equilibrium as two layers. At the higher temperature T_2 the general deviation is less and complete miscibility is indicated, while the inter-

* Aniline and hydrocarbons: Tizard and Marshall, *J. Soc. Chem. Ind.* **40**, 20 T (1921). Phenol and water: Duckett and Patterson, *J. Physical Chem.* **29**, 295 (1925). Glycerol and bases: Parvatiker and McEwen, *J. Chem. Soc.* **125**, 1484 (1925).

mediate temperature T_c of fig. 27b, where a portion of the curve is horizontal, clearly represents an upper limit to the

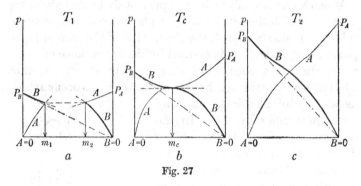

Fig. 27

possibility of the formation of two layers. Fig. 28a shows these effects in terms of composition of the layers and temperature. The temperature T_c, above which complete miscibility prevails, is termed the *critical solution*, or *consolute*, *temperature* (C.S.T.).

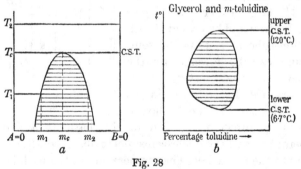

Fig. 28

A limited number of binary liquid systems, especially those formed from an amino- and a hydroxylic-compound, exhibit a second, lower C.S.T., so that complete miscibility is achieved in these cases also by cooling, and a closed curve of mutual solubility is found (fig. 28 b). Although the cause of this behaviour is not at present fully understood, it is thought to be due to a chemical association of the two liquids, leading to *negative* deviations from Raoult's law that offset the more usual positive deviations mentioned above.

When a mixture of two liquids previously heated above the c.s.t., at which the two layers give place to one homogeneous solution, is slowly cooled, the separation of the second liquid phase in small droplets is marked by the appearance of a characteristic opalescence, which disappears on slowly warming. The temperatures at which these opposite effects occur are not usually quite identical, and may sometimes differ by 0·5°. In the construction of a mutual solubility diagram the cooling temperature is considered the more trustworthy.

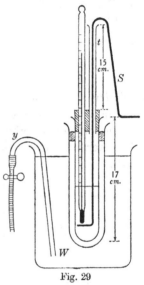

Fig. 29

With the pairs of liquids mentioned below, the apparatus shown in fig. 29 prevents any appreciable change of composition due to volatilization during the observations. The ring stirrer S is of brass wire. The large beaker W serving as water-bath should be provided with a siphon-tube y, so that the temperature can be lowered by the addition of cold water, or, if necessary, ice. The mixture in the inner tube is first made homogeneous by direct immersion in the bath at a suitable temperature, and the opalescence point observed after transferring to the air-jacket. The temperature of the bath is then slightly raised, and the temperature at which the opalescence vanishes is noted. The temperature should not be allowed to fall more than 1° below the opalescence point before the temperature change is reversed.

The proportions suggested in the tables below will be found to give points suitably spaced on the solubility curves. The third column shows the approximate range of bath temperatures required.

(a) $A = methyl\ alcohol$,
   $H = cyclohexane$.

A ml.	H ml.	Temp. °C.
12·5	4·5	0–10
12·5	5·5	15–20
12·5	7·0	25–30
10	10 ⎫	
8	12 ⎪	40–50
6	14 ⎬	
4	16 ⎭	
1	20	20–30
2	20	30–40
3	20	40–50

Sp.gr. (15°): $A = 0·796$,
         $H = 0·784$.

(b) $N = aniline$,
   $P = petrol$, b.p. 80–100°.

N ml.	P ml.	Temp. °C.
19	2	15–20
19	3	25–35
19	4	35–45
16	5	45–50
14	6 ⎫	
12	8 ⎪	
10	10 ⎬	50–60
8	12 ⎪	
6	14 ⎭	
1	19	15–20
2	19	30–40
3	19	40–50
5	19	50–60

Sp.gr. (15°): $N = 1·03$, $P = 0·706$.

For example (b) the aniline should be redistilled to remove water, and be run *slowly* from the burette containing it in order to allow time for the burette to drain.

Recovery of materials. The different experimental mixtures from example (a) are united, and the two layers separated with the aid of a tap funnel. On distilling each layer separately, a constant-boiling mixture, containing 37 % of alcohol, first distils at 54°, and then the remaining pure constituent passes over. The constant-boiling mixture separates into two layers, which may be again distilled. In this way nearly all the materials may be recovered in condition sufficiently pure to be used again.

Plot temperature-composition diagrams (fig. 28), preferably expressing the composition as molar fraction.* In example (b) take the molecular weight of the petrol as that of heptane $C_7H_{16} = 100$. Normally the molar fraction of the mixture showing the C.S.T. is approximately 0·5, as in examples (a) and (b); systems containing water (example (c)) are exceptional.

* For calculation, see p. 78.

(c) *Ph = phenol (as pure crystals), W = water*

Ph g.	W ml.	Temp. °C.
10	4	20–30
	5	40–50
	6	50–60
	8	
	12	
	15	60–70
5	10	
	15	
	25	50–60
2·5	20	40–50
	25	10–20

Warning. Avoid getting concentrated phenol solution on the skin. If this should accidentally happen, wash off immediately in running water.

(d) *Glycerol and* m-*toluidine.*

This system provides a typical example of the closed solubility curve (fig. 28b, p. 87). For anhydrous glycerol the two consolute temperatures are 6·7 and 120°. The lower portion of the curve may be obtained as follows.

Select seven similar test-tubes and provide each with a cork. Label with nos. 1–7 and weigh the tubes separately.

No. of tube	Glycerol ml.	Ratio $\frac{glycerol}{toluidine}$ (by weight)	Bath temp. °C.
1	1	1 : 5	30–40
2	2	1 : 2	
3	4	1 : 1	0–20
4	5	3 : 2	
5	5	2 : 1	
6	6	4 : 1	20–30
7	7	9 : 2	

Attach a paper label to a spare tube and graduate it from
1 to 7 ml. by running water into it from a small burette or
dropping pipette. Warm about 40 ml. of glycerol to about 50°
and pour into the labelled tubes in turn approximately the
volume mentioned in the table above, using the graduated
tube as a guide. Keep the tubes corked to prevent access of
moisture. Re-weigh the tubes to obtain the weight of glycerol,
and then slowly run redistilled m-toluidine ($d = 1·000$) from a
burette into the tubes so as to obtain the weight ratios mentioned in the above table. Set in the tubes in turn a thermometer passing through a cork grooved to admit a small ring
stirrer of brass wire. First warm the mixture to 40–50° while
vigorously stirring to emulsify the base, and then cool at once
in a mixture of ice and water. As the mixture increases in
viscosity reduce the rate of stirring to avoid the introduction
of numerous small air-bubbles. When the mixture has become
clear and homogeneous transfer to an air-jacket immersed in
warm water or tap water as required, and note the temperature of opalescence.

For ordinary 'pure' glycerol ($d = 1·260$) not specially
dehydrated the (minimum) C.S.T. should be at about 14°.

(3) The process of steam distillation

The partial vapour pressures of the constituents of a system
of immiscible liquids will be those exerted by the pure con-
stitutents in the absence of the others. For two constituents
C and W (chlorobenzene and water, see exp. p. 92),

$$(p_C + p_W)_{T°} = (P_C + P_W)_{T°}. \tag{1}$$

Many organic compounds of high boiling-point are sufficiently
nearly immiscible with water for this relation to apply. When
the temperature of such a system is raised, either by direct
heating or by the input of steam, boiling will occur at the
temperature for which $(P_C + P_W)$ = the atmospheric pressure,
and the mixture will distil at this constant temperature in the

molecular proportions P_C/P_W. If the weight ratio of the distillate is G_C/G_W, and the molecular weights M_C and M_W,

$$P_C/P_W = (G_C/M_C) \div (G_W/M_W), \qquad (2)$$

$$M_C = (P_W/P_C) \times (G_C/G_W) \times M_W. \qquad (3)$$

Hence if the quantities on the right-hand side can be determined, as will be shown below, the molecular weight M_C can be found. On rearranging (2) we find

$$G_C/G_W = (P_C \times M_C) \div (P_W \times M_W).$$

The correspondence in magnitude between the products $P \times M$ makes steam distillation a rapid process, even if P_C is low at the temperature of the distillation.

A system suitable for study is formed from water and chlorobenzene, b.p. 132°. The ordinary equipment for steam distillation is required: the distillation flask must, however, be provided with a thermometer carried in the cork and dipping to the middle of the vapour space. The flask should be of not less than 500 ml. capacity.

(i) Calibrate the thermometer for temperatures near 100°, by placing water in the flask, heating first by direct flame and replacing this when the boiling-point is reached by a current of steam, passed through the water at such a rate that distillation is regular. Record the steady temperature registered. (If the thermometer is not graduated a few degrees above 100°, it may be necessary to perform this experiment on a day when the barometer height does not exceed 760 mm.) Ascertain the true boiling-point of water for the barometer height from tables (or assume a change of 1° for a pressure change of 27 mm.).

(ii) Reject the water from the flask, and replace it with a mixture of 150 ml. of chlorobenzene and 50 ml. of water. Again use a direct flame to heat to about 80°, when the steam is passed as before. Regulate the heating of the steam-can so that the distillation proceeds regularly at about 1 drop issuing from the condenser per second. Arrange to collect the mixed distillate in at least two separate vessels. During the course of the distillation record the temperature every minute, and change the receiver when about 50 ml. have been received. Cease the distillation when the temperature rises rapidly (to the boiling-point of water).

(iii) *Analysis of the fractions.* Read the total volume when the whole fraction is placed in a graduated cylinder. Transfer the whole as completely as possible to a tap funnel, and separate the lower (chlorobenzene) layer directly into the cylinder, where the volume is now read. The volume of the water is found by difference. Owing to the great curvature of the interfacial meniscus it is impossible to read each volume accurately when both liquids are together in the cylinder. Greater accuracy can, of course, be secured by weighing the layers, but the further expenditure of time is hardly justified.

Example

Apparent temperature of water distillation = 99·3°.
True boiling-point of water at 752 mm. (barometer) = 99·7°.
Correction to thermometer = + 0·4°.

Apparent temp. of distillation (recorded every minute) °C.	Total distillate ml.	Volume of chloroben- zene ml.	Ratio by volume chloroben- zene to water
(1) 90·6			
90·6			
90·7			
90·7			
90·6			
90·6	44·0	30·0	2·15
(2) 90·7			
90·7			
90·8			
90·8			
90·7			
90·8	62·0	42·5	2·19
Mean 90·7,			Mean 2·17
corrected 91·1°			

Specific gravity of chlorobenzene (at $15°$) = $1·106$.

Ratio of distillate by weight = $2·17 \times 1·106 = 2·37$.

Vapour pressure of water at $91·1° = 548$ mm. (from tables).

Vapour pressure of chlorobenzene by difference = 204 mm.

Using equation (3) $M_C = \dfrac{548}{204} \times 2·37 \times 18 = 114·6$.

(C_6H_5Cl requires $M_C = 112·5$.)

(4) Thermal diagrams

Since the properties of a mixture depend not only on the pressure and temperature, but also on the composition, physical changes, such as melting or freezing, take place not at a fixed temperature but at temperatures that are functions of the composition. Figures showing the relationship between such temperatures and the composition are known as *thermal diagrams*, the description being commonly restricted to diagrams of systems composed of condensed phases, i.e. liquids and solids.

A *cryohydric diagram* is a thermal diagram for a system composed of one or more salts and water. Such systems invariably exhibit eutectic, or cryohydric, points below $0°$, at which three phases, viz. solution, ice and solid salt, are in equilibrium. The term 'cryohydric' survives from a time when the solid part of the eutectic system was mistaken for a 'cryohydrate', the fixity of the temperature and of the composition lending support to this notion. Cryohydric diagrams are generally incomplete on the salt side, as the melting-points of most salts lie far above the highest temperature at which aqueous solutions can exist.

When a substance A is melted and a quantity of substance B is dissolved in the liquid, the properties of the former are altered, and it is very unlikely that the freezing-points of the mixture, i.e. the temperature at which solid first separates on cooling, will be identical with the freezing-point of pure A. Two cases therefore arise, in which the freezing-point of the mixture is (*a*) lower, (*b*) higher, than that of A.

Case (*a*). *The thermal diagram for naphthalene and α-naphthol.* The freezing-point of the homogeneous liquid mixture 1

is shown at T_1 (fig. 30). Unless A and B happen to be homoplastic (for explanation, see p. 36), the stringent conditions
prevailing in crystalline lattices will usually prevent the solid
from containing as much B as the liquid from which it is
being formed; the disparity of composition may be so great
that A separates in a very nearly pure state, as when A = ice.
It follows that as solidification proceeds the liquid becomes
continuously enriched in B, and the temperature of liquid-
solid equilibrium continuously falls, to a temperature T_1',
where solidification is complete. We must distinguish between

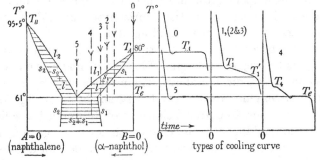

Fig. 30. Naphthalene and α-naphthol.

the *melting*-point at T_1' and the *freezing*-point at T_1; the
temperature range T_1 to T_1' is the *crystallization interval*.

From mixtures in which the proportion of B is further
increased (fig. 30, 2 and 3) we may deduce other pairs of temperatures, and the thermal behaviour of such mixtures is
portrayed in the pair of lines labelled l_1 and s_1 respectively, in
fig. 30. These lines (*liquidus* and *solidus*) may be interpreted
as showing the temperatures at which various liquid mixtures
l_1 are in equilibrium with corresponding solid mixtures
(solutions) s_1.

We may use similar assumptions and arguments in respect
to mixtures formed by the addition of A to liquid B, and so
deduce a second pair of lines labelled l_2 and s_2 originating in
the melting-point T_B of B.

The two sets of lines must necessarily intersect, and so

define a unique temperature (the *eutectic* temperature T_e), at which two solids are simultaneously in equilibrium with one liquid mixture. Downward extensions of the lines l, s below this temperature represent only metastable conditions (realizable to some extent in under-cooled systems, see below).

Melt 10 g. of naphthalene in a large test-tube by immersion for a short time in a boiling water-bath. Transfer to an air-jacket (fig. 20, p. 74), and stir the contents by a gentle but regular rotation of a thermometer dipping into the liquid. Record the temperature at intervals on a cooling curve plotted on squared paper, until the (horizontal) 'arrest' is well established (fig. 30, curve 0, and fig. 31).

Add 2 g. of α-naphthol to the tube, melt the contents *completely*, as before, and well mix. Plot the cooling curve under the conditions indicated above. The arrest is no longer horizontal but extends over a crystallization interval (fig. 31). Owing, however, to extreme slowness of diffusion, the solid, unlike the liquid, cannot undergo continuous change of composition and remain homogeneous, as demanded for a determination of the true lower limit of the interval, i.e. the melting-point. It is not, therefore, worth while to prolong observations on the cooling curve after the direction of the arrest line is well established.

Make further additions of the naphthol as shown in the table, and in each case plot the cooling curve. The last two mixtures should each exhibit arrests in two sections, the latter horizontal and at constant temperature for both mixtures

Naphthalene $(M = 128)$ g.	Portions of α-naphthol added g.	Total naphthol $(M = 144)$ g.
10	2	2
	3	5
	3	8
	2	10

(fig. 30, curve 4). This is the *eutectic temperature* of the system. The two sections are usually separated by a short interval of undercooling.

The diagram may be completed on the naphthol side by carrying out a second series of experiments in which the proportions of the mixtures are reversed.

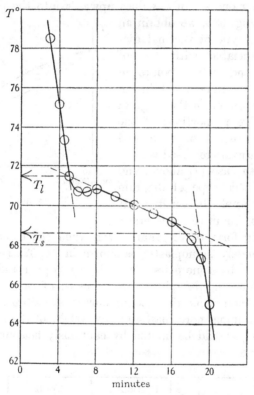

Fig. 31. A typical cooling curve.

To eliminate as far as possible the effect of undercooling the arrest line is projected backwards (fig. 31) and the temperature T_l taken as the true freezing-point. Plot the freezing-points as ordinates against composition (in molar fraction) as abscissae in a thermal diagram, obtaining the composition of the eutectic liquid as the intersection of the freezing-point lines with the horizontal of eutectic temperature. Prepare a mixture in the eutectic proportions indicated and confirm its character by obtaining a single, horizontal arrest on the cooling curve (fig. 30, curve 5).

Case (b). The thermal diagram of naphthalene and β-naphthol. By turning one set of *l, s* lines upwards into the positions shown in fig. 33 *b*, we obtain an important type of thermal diagram associated with *rise* of freezing- and melting-points on one side. It will be noted that in the lower section of the diagram the relative proportions of the two constituents in solid and liquid phases are reversed as compared with case (*a*) above, and therefore this type of diagram can only apply to constituents freely miscible in the solid state.

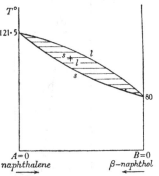

Fig. 32. Naphthalene and β-naphthol.

A limiting form of diagram, given when the constituents are completely homoplastic, is shown in fig. 32. This form is exhibited by some pairs of metals, e.g. copper and nickel. Naphthalene and β-naphthol also show this form.

The method of plotting cooling curves is again used. Mixtures containing more than 40 % by weight of the naphthol (m.p. 121°) should be melted by cautiously heating over a small flame.

Naphthalene ($M = 128$) g.	Portions of β-naphthol added g.	Total naphthol ($M = 144$) g.
10	1	1
	1	2
	3	5
	3	8

The following data are taken from Crompton and Whiteley*
for the α-naphthol system, and from Rheinboldt and Kir-
cheisen† for the β-naphthol system:

Molar fraction α-naphthol	Freezing-point	Molar fraction β-naphthol	Freezing-point	Melting-point
0·00	79·8°	0·00	80·0°	
0·10	74·8	0·132	86·0	80·5
0·20	71·2	0·151	88·0	81·5
0·30	65·7	0·294	93·0	83·5
0·35	64·4	0·346	95·0	84·5
0·395 eutectic 61·0		0·472	99·5	88·5
0·40	62·2	0·569	104·0	93·0
0·50	69·6			
0·60	74·5	0·764	113·5	103·5
0·70	80·8	0·870	117·0	109·5
0·82	87·5	1·00	121°	
0·90	92·5			
1·00	59·5			

('Technical' quality naphthols give quite regular results,
which will, however, not entirely agree with the above data:
the eutectic temperature will be found at about 59°.)

*Examples of the prediction of thermal diagrams by the applica-
tion of the phase rule.* The symbols s_1, s_2 and l will be used to
signify solid and liquid phases in general, and must not be
considered to attach to particular compositions or systems
of phases. In a two-component (or binary) system the rule
($F = C + 2 - P$) demands for non-variance four co-existing
phases, but under the usual practical conditions of open
vessels under (constant) atmospheric pressure, three co-
existing phases suffice. If there is complete miscibility in the
liquid state, and no compounds are formed between the components A and B, the only possible non-variant system will be
s_1, s_2 and l (two solids and one liquid). In this conjunction of
phases two possibilities arise: (a) the composition of l lies

* J. Chem. Soc. **67**, 327 (1895).
† J. prakt. chem. **113**, 202 (1926).

between the compositions of s_1 and s_2; or (b) both solids are richer in one component than the liquid.

Case (a). In this we shall assume, with no loss of real generality, that the components A and B are not appreciably miscible in the solid state, so that s_1 = nearly pure B and s_2 = nearly pure A. Suppose that the non-variant system has been prepared with such proportions of A and B that the gross composition is at C_1 (fig. 33a). This will mean that the mass of the (A-rich) solid s_2 is markedly less than that of either s_1 or l. On adding heat to the system both s_1 and s_2 melt and

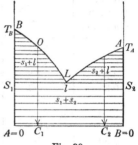

Fig. 33 a

increase the mass of liquid l, which, of course, remains of constant composition as long as both solids are present, since non-variance is maintained. Rise of temperature happens only after all of s_2, the solid present originally in least mass, has liquefied, and the non-variant system has been replaced by the two-phase (univariant) system s_1, l. As the temperature rises s_1 continues to melt, and, consisting of nearly pure B, steadily enriches the liquid in this component. The composition of the liquid therefore inevitably moves to the left, along the (liquidus) line LO. At the temperature of the point O the liquid alone has the original gross composition, and therefore at this point all of s_1 has also melted. The line LO may be carried farther to the left, along LB, by taking a still more B-rich original system, and its final limit is the (unary) system of pure B, melting at T_B.

In a precisely similar way preparations of the non-variant system of gross composition between that of l and s_2 (A-rich) yield the liquidus LA, terminating in the melting-point T_A of A. Finally, if heat is removed from the non-variant system the liquid phase disappears, and the remaining possible two-phase system, viz. s_1, s_2, comes into existence.

It will now appear that the non-variant system is synonymous with a eutectic system; and it follows from the diagram that if in the non-variant system the composition of l is intermediate between those of s_1 and s_2, addition of A to B, or of B to A, lowers the melting-points of B and A respectively. It should be particularly noticed that it is redundant (and not in general true, see case (b) below) to postulate such a lowering of melting-points for the construction of the diagram, but that the application of the phase rule alone, along the above lines, proves the lowering to exist, when the character of the non-variant system is that chosen.

Case (b). For this case to exist, solid solutions, or compounds, must be formed. Suppose first that the gross composition lies at C_1 (fig. 33 b), and from the non-variant system s_1, s_2, l heat is removed. Liquid l freezes to a mixture of s_1 and s_2, and when all of l has solidified, we have the two-phase system s_1, s_2 as before. In general the range of miscibility contracts with fall of temperature, and the conjugate lines diverge, as indicated in the figure. If the gross com-

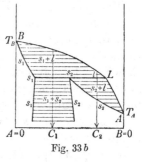

Fig. 33 b

position lies at C_2, then although one phase must disappear on removal of heat, this cannot now be the liquid l, since this can produce a mixture of s_1 and s_2 only if some of the component A is removed from the system. It is a general principle of the construction of diagrams that *a single phase can never give place completely to others unless at least two of these are disposed on opposite sides of the gross composition point.* Actually on the removal of heat it is s_1 that must disappear (it follows from Le Chatelier's principle that s_1 must therefore dissolve with the liberation of heat—a familiar case is that of an anhydrous salt, such as sodium carbonate, dissolving in an aqueous liquid, see fig. 33 c).

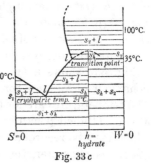

Fig. 33 c

8

After s_1 has completely melted, the two-phase system s_2, l survives, and the temperature falls. The liquid l now progressively solidifies and thereby becomes enriched in A. Consequently l follows the liquidus line LA to the right, and terminates in the melting-point of A at T_A.

On addition of heat to the non-variant system the disappearance (melting) of s_2 takes place, which, in the light of the principle of composition given above, can happen at all gross compositions lying between s_1 and l. The progressive melting of s_1 as the temperature rises in the univariant system s_1, l, enriches l in B, and the liquidus line LB terminates in the melting-point of B at T_B. It follows from the final diagram that if the non-variant system is such as postulated, addition of A to B lowers the melting-point, but addition of B to A raises the melting-point; and this can only happen in a system showing solid solutions. The temperature of the non-variant system corresponds, not to a eutectic temperature, but to a *transition* temperature between the ranges of stability of the solid phases s_1 and s_2. When there is complete miscibility in the solid phase as well as in the liquid, the diagram reduces to that of fig. 32 (p. 98), and no non-variant system is possible (example, Cu, Ni). Under the same conditions fig. 33 a reduces to fig. 33 a', and again no non-variant system exists, but the two-phase system at the minimum of temperature simulates a eutectic system (example, Ni, Cr). Fig. 30 (p. 95) shows the diagram for the intermediate case of solid solutions in a eutectiferous system (example, Cu, Ag).

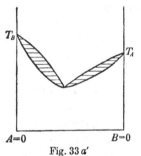

Fig. 33 a'

Fig. 33 c shows a typical diagram (not to scale) for the behaviour of systems composed of water and hydrated salt. The temperatures are for the sodium carbonate-water system already studied (Chapter III, Part 1, p. 72). The diagram, which may be regarded as a combination of cases (a) and (b) above, is characteristic of a large and important class of binary systems in which compound formation occurs. (References to works on the phase rule will be found on p. 52.)

(c) *The cryohydric diagram of potassium nitrate*

Solid solutions cannot be formed with water (ice) and salts, so that on the water side of the eutectic (cryohydric) point pure ice is the solid in equilibrium with all solutions, and on the salt side either the anhydrous salt (as with potassium nitrate) or various hydrates in temperature ranges separated by transition points (see exp. *b* (i), p. 73 and figs. 19 and 33*c*). These systems are thus really simple systems (p. 52), but are taken at this point as examples of a special type of thermal diagram. The curve of equilibrium between ice and solution is sometimes called the 'ice-line', or freezing-point line; the equilibrium curve for salt and solution is identical with the ordinary solubility curve.

Make a standard solution of potassium nitrate by dissolving 20 g. of the salt in 100 ml. of water. The specific gravity of this solution at room temperature $(18°) = 1 \cdot 112$; hence 1 ml. contains $0 \cdot 185$ g. of salt and $0 \cdot 927$ g. of water.

In preparing a freezing bath by mixing ice with brine or solid salt, the following points should be borne in mind:

(1) The temperature of the bath should not be more than 5° lower than the freezing-point to be determined.

(2) The bath must contain much unmelted ice, and the relative quantities of the constituents should be adjusted so that the whole excess of ice remains floating in the solution.

(3) Provided excess of ice remains after well mixing, the temperatures reached by ice-brine mixtures depend only on the concentration of the brine added.

The approximate temperatures of baths prepared in the manner indicated are as follows:

Saturated brine with an equal volume of water, mixed with crushed ice $-4°$ to $-5°$
Saturated brine and crushed ice $-9°$ to $-10°$
Powdered salt mixed with finely crushed ice, in proportions such that both solids remain after mixing $-20°$

For most cryoscopic observations upon aqueous solutions the last bath is far too drastic. In preparing an ice-brine mixture first fill the bath with ice and then pour in the brine so that, after mixing, the solution with floating ice fills the bath.

Temperatures should be observed on a thermometer graduated in $0.1°$, a lens being used to increase the nicety of the readings. If for greater accuracy a Beckmann thermometer is used, it will need resetting at least once (p. 130).

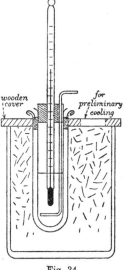

The solution under observation is placed in a wide tube closed with a cork grooved to admit a glass ring stirrer, and through which passes the thermometer. A wider tube to serve as air-jacket (fig. 34) should be supported in the freezing bath (conveniently by the lid of the latter).

(i) *The ice-line.* Immerse the freezing-point tube containing 15 ml. of water directly in the bath (not below $-5°$), and stir continuously until the

Fig. 34

temperature has reached about $-0.5°$; then transfer to the air-jacket, quickly remove the stirrer and take on to it a small fragment of ice (frost) from the outside of the freezing bath, and at once replace in the tube. Ice will now form, and the temperature rise to the freezing-point of water, at which it should remain steady for some time. The thermometer is thus standardized, and the correction determined which must be included in the subsequent estimates of temperature.

Melt the ice by warming in the hand, and then add to the tube 3 ml. of the standard nitrate solution (to form sol. no. 1, table 11), and cool as before, directly in the bath, but to a somewhat lower temperature (say $-2°$), before transferring to the air-jacket and seeding with ice. Water and especially aqueous solutions can readily be undercooled many degrees below the true freezing-point. The weight of ice finally separating will be proportional to the extent of the under-cooling, and unless this is limited (by seeding) to about 0·5°, the concentration of the solution will be materially increased by the withdrawal of solid solvent. The maximum temperature reached when the ice separates will therefore be lower than the temperature of equilibrium between the original solution and ice. To avoid this error, after the approximate freezing-point of a solution has been found, the ice is melted and the determination repeated with at most 0·5° of undercooling.

Reject the solution in the tube, and mix in turn the more concentrated solutions nos. 2 and 3 (table 11), and determine their true freezing-points.

(ii) *The solubility line.* Prepare in the tube solution no. 4, and cool to about 0° by direct immersion in the bath. After placing the tube in the air-jacket introduce a trace of solid nitrate. If this addition causes considerable further deposit of salt, the solution was supersaturated, and its temperature below that of equilibrium between the original solution and nitrate: if the added salt dissolves, the temperature is above the required equilibrium temperature. The temperature must now be adjusted until only a minimal amount of salt is deposited on seeding. The solubility line is very steep in the neighbour-hood of 0°, and it cannot be expected that points on it will be obtained with the same accuracy as those on the ice-line.

Solutions nos. 5–8 inclusive are prepared in turn by succes-sive addition of 1 ml. of water to no. 4. Of these solutions only no. 8 is in stable equilibrium with ice, and is treated like nos. 1, 2 and 3.

(iii) *Metastable conditions.* If solution no. 6, *completely* free from solid, is carefully undercooled, with only *gentle* stirring, to about $-3°$, and then seeded with ice, this solid may separate alone, and a temperature be established on the metastable ex-tension of the ice-line beyond the cryohydric point (fig. 35). A similar experiment may be tried with no. 7. Solution no. 8 is metastable to nitrate, and may be tested in the converse way, by undercooling in the absence of ice to about $-4°$, and then seeding with nitrate. If such experiments are successful the cryohydric composition can be found with exactness by 'bracketing' points so determined, but great care is needed to prevent the cryohydric system from developing spontaneously, and establishing its fixed temperature of $-2\cdot9°$.

Table 11. *Mixtures of water and standard nitrate solution, giving solutions of suitable concentration*

Solution no.	Contents of tube	Quantities to be added		G.-mol. KNO$_3$ per 1000 g. water	Stable solid phase
		Standard solution ml.	Water ml.		
1	Water 15 ml.	3	—	0·312	Ice
2	Water 11 ml.	5	—	0·592	Ice
3	No. 2	5	—	0·914	Ice
4	Standard solution 14 ml.	—	6	1·37	KNO$_3$
5	No. 4	—	1	1·30	KNO$_3$
6	No. 5	—	1	1·24	KNO$_3$
7	No. 6	—	1	1·18	KNO$_3$
8	No. 7	—	1	1·14	Ice

Plot the combined results on squared paper (fig. 35), showing equilibrium temperatures in relation to composition C, expressed as gram-molecules of potassium nitrate ($M = 101$) per 1000 g. of water (calculated values are given in table 11). Draw on the diagram straight lines, joining the freezing-point of water to the points $t = -1\cdot86°$, $C = 1$ and $t = -3\cdot72°$, $C = 1$. These lines are respectively the freezing-point line for an ideal non-electrolyte (N-E), and for an ideal electrolyte (E) with two ions acting independently.

Evaluate from the graph the quantities

$$i = \frac{\Delta t_{\mathrm{KNO_3}}}{\Delta t_{(N\text{-}E)}}$$

(Δt = lowering of freezing-point below that of water) and

$$g = \frac{\Delta t_{\mathrm{KNO_3}}}{\Delta t_{(E)}} = \frac{i}{2},$$

for a series of values of C on the ice-line. $\alpha = i - 1$ gives the 'degree of dissociation' on the Arrhenius theory of electrolytes, and g, *the osmotic coefficient*, the virtual degree of dissociation considered in the modern theories (see p. 176).

(It is of interest to evaluate $\alpha^2/(1-\alpha)V$ (table 12); for this purpose assume the specific gravity of the solutions on the

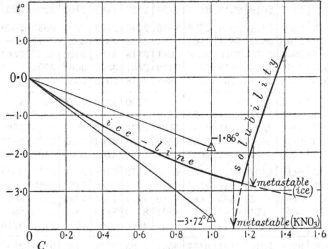

Fig. 35. Cryohydric diagram of potassium nitrate.

ice-line $= 1$. $1/V$ is then the weight concentration; thus solution no. 1, with $31 \cdot 5$ g. $KNO_3/1000$ g. of water, has weight concentration $= 1/V = 0 \cdot 312/1 \cdot 0315$. The error introduced by this approximation does not exceed 5%. Modern theories of electrolytes (see Chapter VI, p. 176) show that as a first approximation $1 - g = k\sqrt{C}$, for uni-univalent salts. To test this, plot g against $\sqrt{C}$, and show that on extrapolation to $C = 0$, $g = 1$.

Table 12. Table of values of $\dfrac{\alpha^2}{1-\alpha}$, for values of α

from 0·0100–0·999

α	0	1	2	3	4	5	6	7	8	9
0·010	1010	1030	1051	1072	1093	1114	1136	1157	1179	1201
0·011	1223	1246	1268	1291	1315	1337	1361	1385	1408	1433
0·012	1457	1482	1507	1532	1557	1582	1608	1633	1659	1686
0·013	1712	1739	1765	1792	1820	1847	1875	1903	1931	1959
0·014	1987	2016	2045	2074	2104	2133	2163	2193	2223	2253
0·015	2284	2314	2345	2376	2408	2440	2473	2505	2537	2569
0·016	2602	2635	2668	2706	2734	2768	2802	2836	2871	2905
0·017	2940	2975	3010	3046	3081	3118	3154	3190	3226	3262
0·018	3299	3336	3373	3411	3449	3487	3525	3563	3602	3641
0·019	3680	3719	3758	3798	3838	3878	3918	3958	3999	4040
0·020	4082	4123	4164	4206	4248	4290	4333	4376	4418	4461
0·021	4505	4548	4591	4635	4680	4724	4759	4813	4858	4903
0·022	4949	4994	5041	5087	5133	5179	5226	5273	5320	5367
0·023	5415	5462	5510	5558	5607	5655	5704	5753	5802	5852
0·024	5902	5952	6002	6052	6103	6154	6204	6256	6307	6358
0·025	6410	6462	6514	6567	6619	6672	6725	6778	6832	6886
0·026	6940	6995	7049	7104	7159	7213	7269	7324	7380	7436
0·027	7492	7548	7605	7662	7719	7777	7834	7892	7949	8007
0·028	8066	8124	8183	8242	8301	8360	8420	8478	8538	8599
0·029	8661	8721	8782	8844	8905	8966	9028	9090	9152	9215
0·030	9278	9341	9404	9467	9531	9595	9659	9723	9788	9852
0·031	9917	9982	1005	1011	1017	1025	1031	1038	1044	1051
0·032	1057	1063	1070	1077	1084	1091	1098	1104	1111	1118
0·033	1125	1132	1138	1146	1153	1160	1167	1174	1181	1188
0·034	1196	1204	1212	1219	1226	1233	1241	1248	1255	1263
0·035	1270	1277	1285	1292	1300	1307	1314	1322	1330	1337
0·036	1345	1352	1360	1368	1375	1383	1391	1398	1406	1414
0·037	1422	1430	1438	1446	1454	1462	1470	1478	1486	1494
0·038	1502	1510	1518	1526	1534	1543	1551	1559	1567	1575
0·039	1583	1592	1600	1608	1616	1625	1633	1642	1650	1658
0·040	1667	1675	1684	1692	1701	1710	1718	1727	1736	1744
0·041	1753	1762	1770	1779	1788	1797	1805	1814	1823	1832
0·042	1841	1850	1859	1868	1877	1886	1895	1904	1913	1922
0·043	1932	1941	1950	1959	1968	1978	1987	1996	2005	2015
0·044	2024	2034	2043	2053	2062	2071	2081	2090	2100	2110
0·045	2119	2129	2139	2149	2159	2168	2178	2188	2198	2208
0·046	2217	2227	2237	2247	2257	2267	2277	2287	2297	2307
0·047	2317	2327	2337	2347	2357	2368	2379	2389	2399	2409
0·048	2420	2430	2440	2450	2461	2471	2482	2492	2503	2513
0·049	2524	2534	2545	2555	2566	2577	2587	2599	2610	2620

α	0	1	2	3	4	5	6	7	8	9
0·050	2631	2642	2653	2663	2674	2685	2696	2707	2718	2729
0·051	2741	2752	2763	2774	2785	2796	2807	2818	2829	2840
0·052	2852	2863	2874	2885	2897	2908	2919	2931	2942	2953
0·053	2965	2977	2989	3000	3012	3023	3035	3047	3058	3070
0·054	3081	3093	3105	3116	3128	3140	3152	3164	3176	3187
0·055	3199	3211	3223	3235	3248	3260	3272	3284	3296	3308
0·056	3321	3333	3345	3357	3370	3383	3395	3407	3419	3432
0·057	3444	3457	3469	3481	3494	3507	3520	3532	3545	3558
0·058	3570	3583	3595	3608	3621	3634	3647	3660	3673	3686
0·059	3699	3711	3724	3737	3751	3764	3777	3790	3803	3816
0·060	3830	3843	3856	3870	3883	3896	3910	3923	3936	3950
0·061	3963	3977	3990	4004	4017	4030	4044	4057	4071	4084
0·062	4098	4111	4125	4139	4153	4166	4180	4194	4208	4222
0·063	4236	4250	4264	4278	4292	4306	4320	4334	4348	4362
0·064	4376	4391	4405	4419	4434	4448	4462	4477	4491	4505
0·065	4519	4534	4548	4563	4577	4592	4606	4621	4635	4650
0·066	4664	4679	4694	4708	4723	4738	4752	4767	4782	4796
0·067	4811	4826	4841	4856	4871	4886	4901	4916	4931	4946
0·068	4961	4976	4992	5007	5023	5038	5054	5069	5085	5100
0·069	5115	5130	5146	5161	5177	5192	5208	5223	5239	5254
0·070	5269	5284	5300	5316	5331	5347	5362	5378	5394	5410
0·071	5426	5442	5458	5474	5490	5506	5522	5538	5554	5570
0·072	5586	5602	5619	5636	5652	5668	5685	5701	5717	5733
0·073	5749	5766	5782	5799	5815	5832	5848	5865	5881	5898
0·074	5914	5931	5947	5964	5981	5997	6014	6031	6047	6064
0·075	6081	6098	6115	6132	6149	6166	6183	6200	6217	6234
0·076	6251	6268	6286	6303	6320	6338	6355	6372	6390	6407
0·077	6424	6442	6459	6477	6494	6512	6529	6547	6564	6582
0·078	6599	6617	6634	6652	6670	6687	6705	6723	6740	6758
0·079	6776	6794	6812	6829	6847	6865	6883	6901	6919	6937
0·080	6955	6973	6992	7010	7029	7047	7066	7084	7103	7121
0·081	7139	7158	7176	7197	7215	7234	7252	7270	7288	7307
0·082	7325	7344	7362	7381	7400	7418	7437	7456	7474	7495
0·083	7513	7532	7551	7570	7589	7608	7627	7646	7665	7684
0·084	7703	7722	7741	7761	7780	7799	7819	7838	7857	7876
0·085	7896	7916	7935	7955	7975	7994	8014	8033	8053	8072
0·086	8092	8112	8131	8151	8171	8190	8210	8230	8250	8270
0·087	8290	8310	8330	8350	8370	8391	8411	8431	8451	8471
0·088	8491	8511	8532	8552	8572	8593	8613	8633	8654	8674
0·089	8695	8715	8736	8757	8777	8798	8819	8839	8860	8881

α	0	1	2	3	4	5	6	7	8	9
0·090	8901	8922	8942	8963	8984	9005	9026	9047	9068	9089
0·091	9110	9131	9152	9173	9195	9216	9237	9258	9280	9301
0·092	9322	9343	9365	9386	9408	9429	9451	9472	9494	9515
0·093	9536	9557	9579	9601	9622	9644	9666	9687	9709	9731
0·094	9753	9775	9796	9818	9840	9862	9884	9906	9928	9950
0·095	9972	9994	1002	1004	1006	1008	1011	1013	1015	1017
0·096	1020	1022	1024	1027	1029	1031	1033	1036	1038	1040
0·097	1042	1044	1047	1049	1051	1054	1056	1058	1060	1063
0·098	1065	1067	1069	1072	1074	1076	1079	1081	1083	1086
0·099	1088	1090	1092	1095	1097	1099	1101	1104	1106	1109
0·10	1111	1135	1159	1183	1207	1232	1257	1282	1308	1333
0·11	1360	1386	1413	1440	1467	1494	1522	1550	1579	1607
0·12	1636	1666	1695	1725	1755	1786	1817	1848	1879	1911
0·13	1943	1975	2007	2040	2073	2107	2141	2175	2209	2244
0·14	2279	2314	2350	2386	2422	2459	2496	2533	2571	2609
0·15	2647	2686	2725	2764	2803	2843	2883	2924	2965	3006
0·16	3048	3090	3132	3174	3217	3261	3304	3348	3392	3437
0·17	3482	3527	3573	3619	3665	3712	3759	3807	3855	3903
0·18	3951	4000	4049	4099	4149	4199	4250	4301	4353	4403
0·19	4457	4509	4562	4616	4670	4724	4778	4833	4888	4944
0·20	5000	5056	5113	5171	5228	5286	5345	5403	5463	5522
0·21	5582	5643	5704	5765	5826	5889	5951	6014	6077	6141
0·22	6205	6270	6335	6400	6466	6532	6599	6666	6734	6802
0·23	6870	6939	7008	7078	7148	7219	7290	7362	7434	7506
0·24	7579	7652	7726	7800	7875	7950	8026	8102	8179	8256
0·25	8333	8411	8490	8569	8648	8728	8809	8890	8971	9053
0·26	9135	9218	9301	9385	9470	9554	9640	9726	9812	9899
0·27	9986	1007	1016	1025	1034	1043	1052	1061	1070	1080
0·28	1089	1099	1108	1117	1127	1136	1146	1155	1165	1175
0·29	1185	1194	1204	1214	1224	1234	1245	1255	1265	1275
0·30	1286	1296	1307	1317	1328	1339	1349	1360	1371	1382
0·31	1393	1404	1415	1426	1437	1449	1460	1471	1483	1494
0·32	1506	1518	1529	1541	1553	1565	1577	1589	1601	1613
0·33	1625	1638	1650	1663	1675	1688	1700	1713	1726	1739
0·34	1752	1765	1778	1791	1804	1817	1831	1844	1857	1871
0·35	1885	1898	1912	1926	1940	1954	1968	1982	1996	2011
0·36	2025	2040	2054	2068	2083	2098	2113	2128	2143	2158
0·37	2173	2188	2203	2219	2234	2250	2266	2281	2297	2313
0·38	2329	2345	2361	2378	2394	2410	2427	2443	2460	2477
0·39	2493	2510	2527	2545	2562	2579	2596	2614	2631	2649

α	0	1	2	3	4	5	6	7	8	9
0·40	2667	2685	2702	2720	2739	2757	2775	2793	2812	2830
0·41	2849	2868	2887	2906	2925	2944	2963	2983	3002	3022
0·42	3041	3061	3081	3101	3121	3141	3162	3182	3203	3223
0·43	3244	3265	3286	3307	3328	3349	3371	3392	3414	3435
0·44	3457	3479	3501	3523	3546	3568	3591	3613	3636	3659
0·45	3682	3705	3728	3752	3775	3799	3822	3846	3870	3894
0·46	3919	3943	3967	3992	4017	4042	4067	4092	4117	4142
0·47	4168	4194	4219	4245	4271	4298	4324	4351	4377	4404
0·48	4431	4458	4485	4512	4540	4568	4595	4613	4651	4680
0·49	4708	4736	4765	4794	4823	4852	4881	4911	4940	4970
0·50	5000	5030	5060	5091	5121	5152	5183	5214	5245	5277
0·51	5308	5340	5372	5404	5436	5469	5501	5534	5567	5600
0·52	5633	5667	5701	5734	5768	5803	5837	5871	5906	5941
0·53	5977	6012	6048	6083	6119	6155	6192	6228	6265	6302
0·54	6339	6377	6414	6452	6490	6528	6566	6605	6644	6683
0·55	6722	6762	6801	6841	6882	6922	6963	7003	7044	7086
0·56	7127	7169	7211	7253	7296	7339	7382	7425	7468	7512
0·57	7556	7600	7645	7689	7734	7779	7825	7871	7917	7963
0·58	8010	8056	8103	8151	8199	8246	8295	8343	8392	8441
0·59	8490	8540	8590	8640	8691	8741	8792	8844	8896	8948
0·60	9000	9053	9106	9159	9213	9267	9321	9375	9430	9485
0·61	9541	9597	9653	9710	9767	9824	9882	9940	9998	1006
0·62	1012	1018	1024	1030	1036	1042	1048	1054	1060	1066
0·63	1073	1079	1085	1092	1098	1105	1111	1118	1124	1131
0·64	1138	1145	1151	1158	1165	1172	1179	1186	1193	1200
0·65	1207	1214	1222	1229	1236	1244	1251	1258	1266	1274
0·66	1281	1289	1297	1304	1312	1320	1328	1336	1344	1352
0·67	1360	1369	1377	1385	1393	1402	1410	1419	1428	1436
0·68	1445	1454	1463	1473	1482	1491	1499	1508	1517	1526
0·69	1536	1545	1555	1564	1574	1583	1593	1603	1613	1623
0·70	1633	1643	1654	1664	1674	1685	1695	1706	1717	1727
0·71	1738	1749	1760	1771	1783	1794	1805	1817	1828	1840
0·72	1851	1863	1875	1887	1899	1911	1924	1936	1949	1961
0·73	1974	1987	1999	2012	2025	2039	2052	2065	2079	2092
0·74	2106	2120	2134	2148	2162	2177	2191	2206	2220	2235
0·75	2250	2265	2280	2296	2311	2327	2342	2358	2374	2390
0·76	2407	2423	2440	2456	2473	2490	2508	2525	2542	2560
0·77	2578	2596	2614	2632	2651	2669	2688	2707	2727	2746
0·78	2766	2785	2805	2825	2846	2866	2887	2908	2929	2950
0·79	2972	2994	3016	3038	3060	3083	3106	3129	3153	3176

α	0	1	2	3	4	5	6	7	8	9
0·80	3200	3224	3249	3273	3298	3323	3348	3374	3400	3427
0·81	3453	3480	3507	3535	3562	3590	3619	3648	3677	3706
0·82	3736	3766	3796	3827	3858	3889	3921	3953	3986	4019
0·83	4052	4086	4120	4155	4190	4225	4262	4298	4335	4372
0·84	4410	4448	4487	4526	4566	4606	4648	4689	4731	4773
0·85	4816	4860	4905	4950	4995	5042	5088	5136	5184	5233
0·86	5283	5333	5384	5436	5489	5542	5597	5652	5708	5765
0·87	5822	5881	5941	6001	6063	6125	6189	6253	6319	6386
0·88	6453	6522	6593	6664	6737	6811	6886	6963	7041	7120
0·89	7201	7283	7367	7453	7540	7629	7719	7812	7906	8002
0·90	8100	8200	8302	8406	8513	8621	8732	8846	8962	9080
0·91	9201	9324	9452	9581	9714	9850	9989	1013	1028	1043
0·92	1058	1074	1090	1107	1123	1141	1158	1177	1196	1215
0·93	1236	1256	1277	1299	1321	1345	1369	1393	1419	1445
0·94	1473	1501	1530	1560	1592	1624	1658	1692	1728	1766
0·95	1805	1846	1888	1933	1979	2027	2077	2130	2185	2244
0·96	2304	2368	2436	2506	2582	2660	2744	2833	2928	3092
0·97	3136	3251	3374	3507	3649	3803	3970	4150	4347	4564
0·98	4802	5005	5358	5684	6052	6468	6945	7493	8134	8892
0·99	9801	1091	1230	1409	1647	1980	2480	3313	4980	9980

Chapter IV

DILUTE SOLUTIONS

PART 1. DETERMINATION OF MOLECULAR WEIGHT IN SOLUTION

The ideal relations (see p. 78)

$$p_A = P_A N_A, \qquad (1)$$

$$p_B = P_B N_B \qquad (2)$$

apply to mixtures of A and B in the ranges of composition respectively represented by $N_A \simeq 1$ (N_B small) for (1), and $N_B \simeq 1$ (N_A small) for (2) (fig. 36a), however marked the

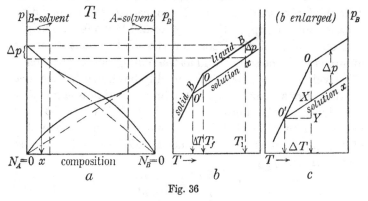

Fig. 36

deviations may be at less disparate values of N_A and N_B. Solutions so approaching ideal behaviour are termed 'dilute', and in them we may logically distinguish the preponderating constituent as the *solvent* (denoted by suffix 1) and the other as the *solute* (suffix 2). By substituting $1 - N_2$ for N_1 in the relations (1) or (2), we obtain

$$(P_1 - p_1)/P_1 = N_2 \ (N_2 \text{ small}). \qquad (3)$$

The quantity on the left of (3) is termed the *relative lowering* of the vapour pressure, and if the solute is non-volatile p_1 is the *total* vapour pressure over the solution. Raoult first announced an empirical generalization of his experimental results in a form closely similar to (3), and for this reason (3) is often referred to as 'Raoult's law', but in this form the law is, in principle, more restricted in scope than (1) or (2), since the relation $1 - N_2 = N_1$ implies that only two molecular species are present in the mixture, and molecular association is excluded. Any theoretical deduction from (3), such as (4) below, primarily refers to solvents which are 'normal' liquids, whose molecular weights in the liquid and vapour states are identical. It should be carefully noted that none of the relations (1), (2) or (3) gives any information about the vapour pressure of the *solute*, which, as fig. 36 shows, is subject to large deviations from Raoult's law. Hence in phenomena in which the vapour pressure (or escaping tendency) of the *solute* is directly concerned, as in partition between solvents (pp. 138 sqq.), Raoult's law offers no help.

If on freezing a dilute solution the solid solvent separates unmixed with the solute, and at the boiling-point the latter is effectively non-volatile, it may readily be shown from simple considerations based on fig. 36 b, c (in which it is assumed that the slopes of the lines are $dp/dT = p \cdot M_1 l / RT^2$), that the freezing-point will be below, and the boiling-point above, that of the pure solvent by an amount $\varDelta T$, given (to a close approximation) by

$$\varDelta T = \frac{RT^2}{M_1 l} N_2 = k \frac{N_2}{M_1}. \qquad (4)$$

l is the appropriate latent heat (fusion or evaporation) per g. and T the freezing- or boiling-point of the pure solvent.

If the solution is highly dilute, we may replace N_2 by the molar ratio $(w_2/w_1)(M_1/M_2)$ and so obtain

$$M_2 = \frac{k}{\varDelta T} \frac{w_2}{w_1}. \qquad (5)$$

Thus from observations on the depression of freezing-point or elevation of boiling-point of solutions of known composition

Table 13. *Freezing-point and boiling-point constants**

$$k = RT^2/l$$

	$k_{f.p.}$	Freez-ing-point		$k_{b.p.}$	Boiling-point
Benzene	5065	5·53°	Acetone	1720	56·2°
Bromoform	14980	8·05	Methyl alcohol	861	64·6
c-Hexane	20810	6·55	Ethyl alcohol	1200	78·3
Nitrobenzene	6864	5·75	Benzene	2633	80·1
Water	1858	0·0	Chloroform	3777	61·2
			Di-ethyl ether	2094	34·6

(w_2/w_1) we can calculate M_2 from (5), provided it is not invalidated by using solutions of too high values of N_2. Equation (5) also affords a means of determining latent heats of fusion and evaporation when solutes of known molecular weights are used.

Associated liquids as solvents

The results of attempts to gain information about the molecular weights of associated or 'abnormal' liquids by treating them as solutes in normal solvents show that they can be divided into two classes, according to the behaviour of the apparent molecular weight with increasing concentration of the solution:

(1) The molecular weight appears to increase steadily without limit throughout the range of concentration in which Raoult's law may be expected to hold. This is typical of the alcohols and phenols (p. 121). Experiments on water are limited by its low solubility (but see exp. *a*, p. 134, on water dissolved in nitrobenzene).

(2) The molecular weight remains constant at double the value for the simplest chemical formula. This is characteristic of carboxylic acids (p. 120). The dimeric association sometimes persists in the vapour state (as in acetic acid).

* Data for T and l from Timmermans, *op. cit.* on p. 8 and *Natl. Bur. Stand, op. cit.* on p. 8.

When employed as cryoscopic or ebullioscopic solvents abnormal liquids show surprisingly little trace of the molecular complexity suggested by the above results. Raoult and his successors* have shown that the results of direct measurement of the relative lowering of the vapour pressure of aqueous solutions can be reproduced by equation (3) if the molar fraction is calculated by taking $M_1 = 18$. It may, however, be noted as an experimental fact that although water and the alcohols usually give quite regular results when (5) is applied to them acting as solvents, the apparent molecular weights of solutes are very often the lowest possible consistent with the structural formula, even when marked association is indicated in other types of solvent. In the present state of knowledge it is probably unwise to regard normal liquids as 'causing' association in such cases; they are better classed as inert solvents.

As the relation (5) becomes more exact as ΔT becomes smaller, determinations of molecular weight based upon it necessarily demand a thermometer accurately registering small differences of temperature. In all the examples suggested in the following pages reasonably satisfactory results can be obtained if a lens is used to observe temperatures on a thermometer graduated in 0·1°, and for large classes such instruments will no doubt be the best available. A description of manipulation with the more delicate Beckmann thermometer will be found at the end of the section (p. 130).

Cryoscopic experiments

A. *Solutions in water*

The apparatus and method described in exp. *c*, p. 103, are suitable for general use. The precautions regarding temperature of the freezing-bath and undercooling there detailed should be noted. For rapidly cooling the solvent at the beginning, or the solution after an addition of solute, it is very convenient to have at the bottom of the air-jacket a small pool of mercury, into which the freezing-point tube can be depressed, to be

* E.g. Pearce and Snow, *J. Physical Chem.* **31**, 231 (1927).

raised again and 'seeded' when a temperature in general not more than 0·3° below the true freezing-point has been reached. This cooling device saves much time by making it unnecessary to remove the inner tube during a series of observations, and also allows the undercooling to be accurately judged.

(a) *Confirm equation* (5) *for water, and evaluate the constant* k_f, *by using solutions of cane-sugar, mannitol, or urea.*

For 20 ml. of water, add in all about 6 g. of cane-sugar, 1·5 g. of urea, or not more than 2·0 g. of mannitol, in three approximately equal portions.

Plot on squared paper the freezing-points of the water and the solutions against weight of solute w_2. Draw the best straight line through the points, and estimate its slope $\Delta T/w_2$, i.e. the lowering for 1 g. of solute. Then from equation (5)

$$k = M_2 . w_1 . \text{(slope of line)}.$$

$$C_{12}H_{22}O_{11} = 342; \ C_6H_{14}O_6 = 182; \ CON_2H_4 = 60.$$

(b) *The formation of complex ions by the mercuric halides.*[*]

Place in the inner tube 20 ml. of water and observe the freezing-point. Add and dissolve 1/50 g.-mol. = 3·3 g. of potassium iodide ($M = 166$), thus obtaining a normal solution of the salt. Again observe the freezing-point (below −3°). Observe freezing-points of the (pale yellow) solutions formed by the successive addition of four portions of red mercuric iodide, each weighing about 1·5 g. The freezing-point tube should be withdrawn from the apparatus and well shaken to dissolve the mercuric iodide, which owing to its high density tends to settle at the bottom and elude the action of the stirrer. The last portion of the mercuric iodide will probably not all dissolve, the solution thus becoming saturated with this substance, but the suspended solid in no way interferes with the freezing-point observation.

Plot the freezing-points of the solutions (which will be found to rise steadily with the addition of mercuric iodide)

[*] Sherill, *Z. physikal. Chem.* **43**, 705 (1903); *ibid.* **47**, 103 (1904); Lindgren *et al.*, *Acta chem. Scand.* **1**, 479 (1947) Strocchi, *Gazzetta*, **79**, 270 (1949).

against the weight of mercuric iodide dissolved (fig. 37). Omit
the last portion, if it did not all dissolve (see below). Draw a
horizontal line MX at the (maximum) freezing-point found
for the final saturated solution. Draw a smooth curve through

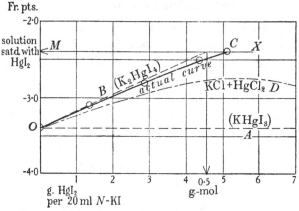

Fig. 37. Effect of mercuric iodide upon the freezing-point
of a solution of potassium iodide.

the freezing-points and extend it to cut the line MX in C,
which thus gives on the weight axis the weight of HgI_2
required for saturation. Mark on the weight axis the value
4·54 g. = 0·01 g.-mol. HgI_2.

The solvent action of potassium iodide solutions upon the
otherwise quite insoluble mercuric compound can only be explained by chemical union of the two iodides and consequent
formation of stable complex ions:

$$K^+ +\ I^- + HgI_2 \rightarrow\ K^+ + (HgI_3^-), \qquad (1)$$

$$2K^+ + 2I^- + HgI_2 \rightarrow 2K^+ + (HgI_4^=). \qquad (2)$$

If the reaction takes the course (1), the freezing-point of the
KI solution will be almost unaffected by the dissolution of
HgI_2 (since the *number* of ions present is unaltered) and the
solubility of the HgI_2 will reach 1 g.-mol. HgI_2/g.-mol. KI
(line OA). If the reaction is confined to (2), the freezing-point
must rise, as two ions of I^- give place to one complex ion, and
the maximum freezing-point, i.e. that of the saturated solution,

will correspond to a solubility of 0·5 g.-mol. HgI_2/g.-mol. KI (line OB). The actual freezing-point line will be found in position OC, and its contiguity to OB shows that the principal product is in fact the ion HgI_4^-, but to account for the slight deviation towards OA, and a solubility (point C) greater than 0·5 g.-mol./g.-mol. KI, the concurrent formation of small amounts of HgI_3^- may be assumed.

Similar behaviour is shown by $HgBr_2$, $Hg(CN)_2$ and $HgCl_2$ towards the corresponding alkali salts, but the last case (line OD) is complicated by the appreciable solubility of $HgCl_2$ as such, and the considerable dissociation of the complex ions containing chlorine.

B. *Solutions in organic solvents*

All solutes used with organic solvents must be freed from moisture, by exposing them to warm air, or by keeping for some time in a vacuum desiccator, if solids; and by other suitable means if liquids. It should be borne in mind that as the latent and specific heats of organic liquids are usually low, it is especially necessary when using them to limit the undercooling, and to adjust the bath temperature carefully, preferably by the use of a second (common) thermometer.*

A volume of solvent sufficient to immerse the top of the thermometer bulb about 1 cm. below the surface should be run into the freezing-point tube from a burette, which is accurately read. An equal or convenient lesser volume of solvent is at once run from the burette into a weighing bottle, which is then stoppered and weighed. As organic liquids have usually high thermal expansions, considerable inaccuracy may be introduced by estimating the weight of solvent from the density at a fixed temperature.

For seeding, a finely drawn-out tube should be immersed in a test-tube containing a few millilitres of the solvent and kept in the cooling bath. The thermometer and bung should be raised from the freezing-point tube, and the *surface* of the

* Bury and Jenkins, *J. Chem. Soc.* 1934, p. 688.

undercooled solution touched with the tip of the drawn-out tube. On again closing the tube and *gently* stirring, solid should separate and the temperature rise to the freezing-point. Solvents vary greatly in the ease with which they can be undercooled (water > benzene > bromoform), but in general solutions are more readily undercooled than the pure solvents, so that while seeding is often not required for very dilute solutions in bromoform, it becomes necessary for stronger solutions. The rate of rise of temperature after seeding also varies very much and depends both on the latent heat of fusion and on the natural rate of crystallization. The rise is very rapid in water, slower in benzene, and markedly sluggish in bromoform (especially if it contains traces of alcohol, as it usually does). An adequate period of gentle stirring must therefore elapse before the maximum temperature is reached, and if this temperature is to be taken as that of true equilibrium, i.e. the freezing-point, it is obvious that the bath should differ only slightly in temperature from the freezing-point. To melt the solid for a second observation, withdraw the tube to the top of the air-jacket, so that its contents are partially exposed to the warmth of the room; warming in the hand may result in an unnecessarily large rise of temperature, with loss of time in recooling. The *whole* of the solid should have disappeared before cooling is again applied.

At least three concordant observations should be obtained before further solute is added, and it is always advisable to plot on squared paper the freezing-points against weight of solute as the work proceeds, so that irregularities, if due to experimental error, may be more readily brought to light and corrected.

(a) *The cryoscopic behaviour of carboxylic acids in benzene (or bromoform).*

Suitable solid acids are phenylacetic ($C_6H_5CH_2COOH = 136$), or benzoic ($C_6H_5COOH = 122$), but experiments with the latter are limited by its low solubility (eutectic point at 5·35 % acid). A mixture of water and ice only is to be used as freezing bath. First observe and confirm the freezing-point of the pure solvent (20 ml.) and then add from a weighing tube and dissolve about 2 g. in all of the acid in at least four portions of approximately the weights shown in the example below. (Benzoic acid dissolves rather slowly.) Observe and confirm the freezing-point after each addition.

Plot on squared paper ΔT against weight of acid dissolved, when a straight line should be found, showing that equation (5) is obeyed and M_2 is constant. Calculate M_2 from the slope of the line $(\Delta T/w_2)$ by equation (5), p. 114:

$$M_2 = (k/\Delta T)(w_2/w_1) = k/w_1 \text{ (slope)}.$$

Example

Volume of benzene $= 20$ ml. $= 17\cdot60$ g. Freezing-point $= 5\cdot48°$.

Phenyl acetic acid dissolved (g.) (w_2)	Freezing-point	ΔT
0·225	5·23°	0·25°
0·327	5·08	0·40
0·846	4·55	0·93
1·272	4·10	1·38
1·776	3·57	1·91
2·122	3·22	2·26

k_f for benzene from table $13 = 5065$.
Depression for 1 g. of acid $(\Delta T/w_2) = 1\cdot07°$.

$$M_2 = \frac{5065}{17\cdot60 \times 1\cdot07} = 269.$$

($C_6H_5CH_2COOH$ requires $M = 136$.)

(b) *The cryoscopic behaviour of alcohols in benzene (or bromoform).*

A suitable solute is benzyl alcohol ($C_6H_5CH_2OH = 108$), which should be redistilled to remove water (b.p. 206°). The solute is weighed in a small pyknometer (fig. 38), which should have a fine capillary outlet at a, so that a limited number of drops of the liquid can readily be delivered into the solvent by gently blowing on the rubber tube b. In order to be able to judge the number of drops required in one addition it is well to make a preliminary weighing of 5–10 drops received in a test-tube.

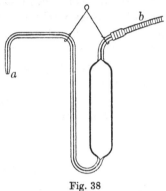

For 20 ml. of benzene or bromoform use approximately the weights shown in the example, or a selection of them.

On plotting the freezing-points against weight of alcohol added a highly curved line is found, of steadily decreasing slope, indicating that the apparent molecular weight is increasing with concentration. Calculate values of the apparent molecular weight from equation (5), as shown in the example.

Fig. 38

Example

Volume of benzene = 20 ml. = 17·60 g.

$$M_2 = \frac{k}{\varDelta T}\frac{w_2}{w_1} = \frac{k}{w_1}\frac{w_2}{\varDelta T}.$$

$$\frac{k}{w_1} = \frac{5065}{17\cdot60} = 288.$$

Benzyl alcohol dissolved (g.) (w_2)	$\varDelta T$	$\dfrac{w_2}{\varDelta T}$	M_2
0·103	0·27°	0·382	109
0·177	0·47	0·377	108
0·308	0·71	0·434	125
0·460	0·97	0·475	136
0·715	1·33	0·538	155
1·031	1·68	0·612	176
1·352	1·98	0·683	197
1·894	2·43	0·780	225

The boiling-point method

When the bulb of a thermometer is placed in the *vapour* of a boiling liquid, as is usually done to observe the boiling-point, it registers the temperature of a true equilibrium between a film of condensed liquid and the vapour, and this temperature is by definition the true boiling-point. The presence in the liquid of non-volatile solute will not alter the observed temperature. To obtain the boiling-point of a solution the bulb of the thermometer must be in contact with the boiling solution and not with condensed vapour.

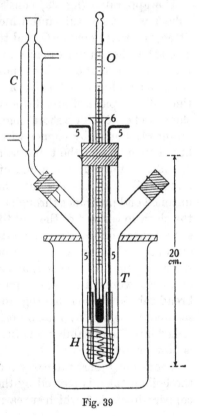

It has proved difficult to obtain the true equilibrium conditions by simply immersing the thermometer bulb in the boiling liquid, for unless this is permeated by a continuous supply of vapour in the immediate vicinity of the bulb, superheating will usually occur and the true equilibrium temperature will not be observed.

Very many forms of apparatus have been devised to overcome this serious difficulty, but it is probable that only in the apparatus of Cottrell* and of the type first suggested by Landsberger†

Fig. 39

* Cottrell, *J. Amer. Chem. Soc.* **41**, 721 (1919); Washburn and Read, *J. Amer. Chem. Soc.* **41**, 739 (1919).

† *Ber.* **31**, 458 (1898); Walker and Lumsden, *J. Chem. Soc.* **73**, 502 (1898).

is it in practice eliminated. In the latter apparatus, in which the solution is heated solely by bubbling a supply of solvent vapour through it, other special difficulties arise. Undoubtedly the best means of ensuring an adequate supply of vapour bubbles is by internal electrical heating. A simple and easily constructed apparatus which combines the advantages of this means of controlled heating with the principle of the Cotrell 'vapour pump' is described below.

The apparatus (fig. 39) consists of three parts: a boiling tube T with a side tube into which a small water condenser is fitted, the thermometer O, and the heating unit H. A second side tube, though not essential, is convenient for introducing the solute without detaching the condenser. The heating unit is made as follows. A *small* hole h (about 1 mm.) is blown in the wall of a piece of broad thin-walled tubing in the position shown in fig. 40 a. Two short lengths of quill tubing (4, fig. 40 b) are firmly bound by copper wires 3 on opposite sides of the broad tube, one of the tubes being contiguous to the hole h. A coil 2 is formed at the end of a length of nichrome wire (s.w.g. 26), and is inserted into the broad tube so that the long uncoiled end protrudes through h (fig. 40 b). By now brazing* the short lower end of the coil to a copper wire of about the same gauge and inserting the latter into the tube 4, the lower end of the coil is secured in position. The projecting long end of the nichrome wire is passed under and once round the tube 4 and then wound in a tight spiral round the outside of the broad tube and side tubes (fig. 40 c). The outside coil is finally secured as before, by brazing on a copper wire, which is inserted into the remaining side tube 4. The lower copper wire 3 is now removed.

A cork bung, bored to carry the thermometer, and fitted to the boiling tube, is pierced by the pointed ends of two stout copper wire leads 5, which are carried down to the heating unit

---

* Twist the copper wire round the nichrome, moisten the join, and cover it with a mixture of brass filings and borax in about equal bulks: heat in a small blowpipe flame till the brass filings melt and flow freely.

and firmly thrust into the side tubes 4 beside the smaller lead
wires from the coils; the small leads are then twisted round
the stouter and their ends soldered to the latter in a position
somewhat above their exit from the side tubes 4. The heating
unit is thus securely carried on the bung, and when the latter
is in position in the boiling tube the unit should rest on the
bottom of the tube, final adjustment being made by sliding
the leads 5 in or out of the side tubes.

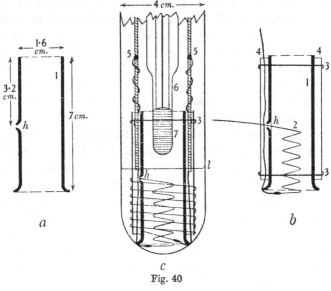

Fig. 40

The heating current (alternating or direct) of about 2 amp.
is most easily obtained from town supply by using carbon
lamps (32 c.p.) in parallel and a small variable resistance
(10 ohms). When in use the boiling tube is shielded from
draughts by inserting it into a beaker through a lid of asbestos
board (fig. 39). If a Beckmann thermometer is to be used the
bung should be originally bored for a wide tube, the lower end
of which is seen in fig. 40 c at 6. The thermometer stands in this
tube with its bulb in the small pool of mercury 7; this arrangement allows the thermometer to be removed for resetting
if necessary, without interrupting the boiling. It is also to be

recommended for its steadying influence on the temperature reading.

The solvent is run into the boiling tube so that its surface (l, fig. 40 c) is just above the top of the coils and somewhat below the thermometer bulb. With the dimensions suggested in the figure about 35 ml. will be required. The liquid is then brought to steady boiling by the heating coils, when the thermometer bulb remains for its greater part in the vapour. The temperature is observed at intervals until it has become steady (10–15 min. from cold). The heating current is then switched off, and, after the vapour has subsided, a weighed amount of solute is introduced, either by temporarily withdrawing the condenser or by the second side-neck. The heating is then resumed, and the liquid vigorously boiled until all solid has been washed down from the side-neck, and the solution has become homogeneous. Owing to the fact that nearly all organic solutes lower the surface tension of solvents, the solution will usually foam well up the boiling tube during this heating. The current is now adjusted by means of the resistance so that the thermometer bulb is enveloped in a continuous stream of foam rising in the broad tube of the heating unit. The temperature rapidly becomes constant, and should remain so even when the rate of boiling is somewhat increased. Electrolysis is quite negligible in an organic solvent even when the solute is an organic acid, but the method of heating cannot be used for aqueous solutions.

(a) Using acetone as solvent, plot the boiling-points of solutions of one of the solutes listed in table 15, p. 130, against weight of solute added, and confirm by obtaining a straight line the validity of equation (5) for the boiling-point elevation.

(b) Make similar observations, with the same amount of solvent, of the boiling-points of solutions of benzoic acid in acetone, using not more than 10 g. of the acid in all per 100 ml. of solvent. Plot the boiling-points as above, when a straight line is again found. Read off from the graphs the respective weights of solute required for a rise of 1°, and then by assuming the molecular weight of the normal solute calculate that of benzoic acid in acetone solution by simple proportion. Compare the effect of acetone on the molecular weight of the acid with that of benzene in exp. *a*, p. 120. Calculate also from the rise shown by the normal solute the constant of elevation for acetone under the conditions of the experiments, and compare the value found with the theoretical value of table 13, p. 115. (The change of theoretical constant with change of barometer pressure is of the order of less than 1 % for 10 mm. and is therefore negligible in ordinary work.)

Choice of solvent for determination of molecular weight

(1) The solute must be sufficiently soluble for a reasonable range of concentration to be studied, and in the cryoscopic method it must be known (or ascertained) that solid solutions are not formed with the solvent (thiophene, phenol, iodine and probably pyridine all form solid solutions with benzene). The eutectic point should not be exceeded.

(2) The apparent advantage of employing solvents with high constants in the cryoscopic method (e.g. cyclohexane or bromoform) is to some extent illusory, as the low latent heat of fusion implied by the high constant (equation (4), p. 114) demands a strict control of the bath temperature.

(3) It may generally be assumed that a molecular weight determined from the boiling-points of solutions in alcohols or acetone will have the minimal value consistent with the simplest chemical formula. If a Beckmann thermometer is used (to counteract the low constant), the lower boiling-point of methyl alcohol and consequent diminished heat losses give it some advantage over ethyl alcohol in the boiling-point method. In 'normal' liquids (hydrocarbons, chloroform or bromoform) a very large number of solutes appear to have molecular weights higher than the minimal.

(4) As table 14 shows, molar fractions run roughly parallel to *volume* (but not *weight*) concentration, and hence comparable conditions, i.e. similar molar fractions, are secured for

experiments with the same solute in different solvents when approximately the same *weights* are dissolved in equal *volumes* of the solvents.

Table 14

Solvent	1000 ml. weigh	G.-mol. solvent per 1000 ml.	Molar fraction of 1 g.-mol. solute per 1000 ml.	G.-mol. solvent per 1000 g.	Molar fraction of 1 g.-mol. solute per 1000 g.
Acetone	790	13·6	0·069	17·3	0·055
Benzene	880	11·3	0·081	12·8	0·072
Chloroform	1500	12·6	0·073	8·4	0·107
Bromoform	2700	10·7	0·085	3·9	0·204

Purification of solvents

When pure (e.g. *Analar* brand) solvents are not available, the following methods may be used:

Benzene (b.p.$_{760}$ 80·10°, f.p. 5·5°). Thiophene is removed as follows:* 1000 ml. of commercial benzene are boiled under reflux for some hours with a solution of mercuric acetate, made by dissolving 40 g. of *freshly precipitated* mercuric oxide in a mixture of 300 ml. of water with 40 ml. of glacial acetic acid. If the liquid is stirred the time of boiling can be shortened to half an hour. Distil off the benzene, dry over potassium hydroxide, and finally distil through a still-head to remove toluene. Alternatively, the dried hydrocarbon may be freed from toluene by partial freezing.

Chloroform (b.p.$_{760}$ 61·2°). 'B.P.' chloroform contains 99–99·4 % of pure chloroform with 1–0·6 % of ethyl alcohol. The simplest way to remove the alcohol and at the same time to dry the solvent is to shake with a small volume of concentrated sulphuric acid. Distil after separating. The solvent, when freed from alcohol, should be kept in the dark to avoid the photochemical formation of phosgene.

* Dimroth, *Ber.* **32**, 759 (1899).

Bromoform (f.p. 8·0°). 'B.P.' material contains 96 % of pure bromoform and 4 % of ethyl alcohol. This should be removed by several washings with water, after which the bromoform is separated and dried over anhydrous sodium sulphate. Final purification is best accomplished by partial freezing with an ice-water bath, or by distillation under reduced pressure. When freed from alcohol the solvent readily undergoes photochemical changes and should therefore be stored in the dark.

Acetone (b.p.$_{760}$ 56·2°). This solvent is rather troublesome to purify. The best method is to distil over solid potassium permanganate, which is without action upon pure acetone. It may be further purified by conversion into the bisulphite compound, filtering and treating the compound with dilute sodium carbonate. Acetone may be dried with anhydrous sodium sulphate, or by standing over quicklime.

Methyl and ethyl alcohols. Both these solvents are now readily obtainable in pure anhydrous condition, in which state, however, they are hygroscopic. For this reason rectified spirit may be preferred to absolute ethyl alcohol, but it is essential that the composition should be that of the constant boiling mixture (96·5 % alcohol). As methyl alcohol does not form a constant boiling mixture with water, it must be used anhydrous, in which condition it may if necessary be prepared by simple fractionation.

For solvents of assured purity the calculated constants k_f of table 13 may be used with confidence in determining molecular weight by the cryoscopic method; but if any doubt exists about the purity of a solvent, its actual constant should be regarded as empirical and determined by using a standard solute (table 15). In employing the boiling-point method the constant k of equation (5) should *always* be treated as empirical (exp. *b*, p. 126), as its apparent value, even for the same solvent and solute, may vary considerably from one apparatus to another, owing to different physical conditions—of heat loss, of superheating, and of amount of vaporized solvent. Owing to its freedom from such uncertainties the cryoscopic method is to be preferred where possible.

The solutes of table 15 are all readily purified by the ordinary methods of organic chemistry, and have been chosen from among those substances generally accepted as showing 'normal' behaviour in the solvents indicated, i.e. exhibiting in these solvents (but not necessarily in others) a constant and minimal molecular weight. In using these solutes as standards the total weight (g.) to be used per 100 *ml.* of solvent shown in the table should not be exceeded, and should generally be added in at least four portions.

Table 15. *Chart of standard solutes*

Solute	M	Freezing-point			Boiling-point			
		Water	Benzene	Bromoform	Methyl or ethyl alcohol	Acetone	Chloroform	Benzene
Naphthalene	128	—	10	10	—	—	10	15
Anthracene	178	—	—	—	—	—	—	6
Triphenylmethane	244	—	—	—	14	12	10	10
Sucrose	342	20	—	—	—	—	—	—
Mannitol	182	10	—	—	—	—	—	—
Resorcinol	110	—	—	—	10	5	—	—
Benzil	210	—	—	—	10	8	7	10
Camphor	152	—	20	10	—	10	12	—
Benzanilide	197	—	—	—	12	10	8	—

The use of the Beckmann thermometer

Since relatively small differences of temperature have to be measured, a scale comprising a range of only 6° C. suffices. In order to be able to use such a thermometer for different solvents, mercury must be added to or removed from its bulb B (fig. 41). For this purpose the upper storage bulb b is brought into use.

Invert the thermometer, and by tapping its upper end bring the mercury in b into connexion with the capillary. Warm the bulb B with the hand, or *very* carefully high above a small flame, until the mercury in the capillary rises to the upper bulb and unites with the drop of mercury therein. Now cool B in a water-bath whose temperature has been adjusted by suitable addition of ice to 2–3° (dependent on the length of capillary between the scale S and b) above the freezing-point of the pure solvent. It is essential to use a second (ordinary) thermometer in adjusting the bath temperature. After allowing sufficient time for the mercury in B to take the bath temperature, tap the top of the bulb as before and so separate the thread from the mercury remaining in the upper bulb. Test the setting by immersing the bulb of the thermometer in the freezing solvent, when the mercury should stand just below the uppermost graduation on the scale S.

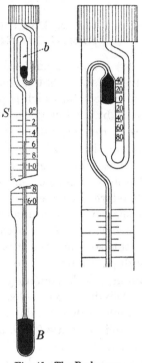

Fig. 41. The Beckmann thermometer.

To set the thermometer for a boiling-point experiment a similar procedure is adopted, but the temperature of the water-bath must be 6–7° higher than the boiling-point of the pure solvent, since in such experiments the mercury must stand at the *bottom* of the scale S when the bulb is at the boiling temperature of the pure solvent. The operation of setting the thermometer is greatly simplified when the upper bulb is graduated, as in fig. 41. It is then only necessary to warm the bulb B until the mercury in b registers the desired setting, and then break the thread by tapping.

The actual registration on the thermometer must vary with the content of mercury in B, but the correction may be neglected between 0 and 10°, in which range the freezing-points of the commoner solvents lie. When used for the boiling-point method the correction is immaterial, as the experiments must usually be made on an empirical basis (as in exp. b, p. 126).

DILUTE SOLUTIONS (*cont.*)

PART 2. EXPERIMENTS ILLUSTRATING
SEMI-PERMEABILITY AND OSMOSIS

(*a*) When a liquid and its vapour are in equilibrium, molecules escape from the liquid and condense from the vapour at the same rate, which is proportional to the vapour pressure, since, by the kinetic theory, the rate of condensation must be proportional to this pressure (cf. introductory notes, p. 53). It follows that any influence which reduces the vapour pressure, such as a fall of temperature, or the presence in the liquid of dissolved substance, must also reduce the rate of escape (and therefore of condensation). If, therefore, an aqueous solution is shaken with an organic solvent, such as nitrobenzene, in which water is to some extent soluble, less water will dissolve in this solvent than if it had been brought into equilibrium with pure water; in short, the presence of dissolved substance diminishes the solubility of the water in the organic solvent.

Dehydrate redistilled nitrobenzene by shaking it with granular calcium chloride in a glass-stoppered bottle. Allow the solid to settle (preferably overnight), leaving above it the clear nitrobenzene, which can be easily decanted as required. Prepare a saturated solution of calcium nitrate by mixing 200 g. of the crystals with 100 ml. of water (this solution will also be needed for exp. *b*). Shake about 50 ml. of the nitrate solution with about 30 ml. of the dry nitrobenzene in a tap funnel for at least 1 min. While allowing time for the layers to separate, determine the freezing-point (about 5·6°) of a specimen of the dry nitrobenzene, by the methods and apparatus previously described in the cryoscopic experiments (pp. 119 sqq.). Then allow the (lower) nitrate solution to flow out of the funnel as completely as possible. Pour the nitrobenzene *from the top of the funnel* into a second freezing-point tube through a pad of glass-wool, which will retain any droplets of the aqueous solution that have failed to separate. On cooling the nitrobenzene so treated it will be noted that the liquid remains clear right down to its freezing-point, which should be carefully determined (seeding may be necessary).

A second specimen of the nitrobenzene is treated in the same way with distilled water, the temperature of the mixture (room temperature) being noted. The nitrobenzene, now the *lower* layer in the funnel, may be run directly out, by means of the tap, through a pad of glass-wool, as before. On cooling this specimen it at once becomes cloudy, owing to separation of

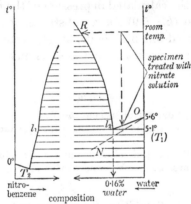

Fig. 42. Nitrobenzene and water.

water from the solution saturated at room temperature. The opacity increases steadily down to the freezing-point (about 5°), which is determined. Owing to the marked undercooling seeding must be used.

Plot on squared paper a curve of the solubility of water in nitrobenzene, using the data in column 2 of table 16 below (fig. 42, curve Rl_2). Read off the solubility (*a*) at the temperature of making the nitrobenzene-water mixture, (*b*) at the freezing-point of this preparation. On a second and larger scale diagram, show freezing-point against percentage of water, by joining the point (*b*) above and the freezing-point of the dry nitrobenzene with a straight line (Ol_2, fig. 42). From this graph read off the percentage of water contained in the nitrobenzene treated with the nitrate solution.

Fig. 42 shows the basis of the above experiments. From all mixtures of nitrobenzene and water containing 0·16 % or more of water in the nitrobenzene, solid nitrobenzene separates at the temperature T_1, where the solid is in equilibrium with two mutual solutions l_1 and l_2. The system at this temperature is analogous to a cryohydric system, which also has three phases (two solids and one solution): such a condition is found at T_2, where ice, solid nitrobenzene, and an aqueous solution of the latter are in equilibrium. The dotted line ON shows the calculated depression of the freezing-point of nitrobenzene ($k_f = 6·9$) for a substance of mol. wt. = 18. Since the observed line lies above, at Ol_2, water is somewhat associated in nitrobenzene solution, as might have been expected; cf. p. 115.

Example

Freezing-point of dry nitrobenzene = 5·60°.
Freezing-point of specimen treated with calcium nitrate solution = 5·25°.
Temperature of mixing nitrobenzene and water = 18·0°.
Solubility of water in nitrobenzene at this temperature = 0·215 % (table 16).

Table 16. *Solubility of water in nitrobenzene*

Temp. °C.	% water
10	0·18
20	0·22
30	0·27
40	0·365
50	0·47

Solubility of nitrobenzene in water

Temp. °C.	% nitrobenzene
20	0·19
30	0·22
55	0·27

Freezing-point of nitrobenzene saturated with water = 5·12°.
Solubility of water in nitrobenzene at 5·12° = 0·16 %.
Solubility of water from nitrate solution in nitrobenzene (from graph) = 0·115 %.
Ratio of solubilities water/nitrate solution = 0·215/0·115 = 1·87.
(From this result it follows that the vapour pressure over the nitrate solution at 18° is (vapour tension of water) ÷ 1·87 = 15·3 ÷ 1·87 = 8·2 mm.)

(b) From the facts illustrated by the above experiments, it must follow that if a solution is separated from solvent by a septum of material X (fig. 43), in which the solvent, but not the solute, is somewhat soluble, there will be a concentration gradient of dissolved solvent across the septum, down which solvent must continuously flow into the solution. This is the process of *osmosis*. A septum which thus allows the passage through itself of only one constituent of a solution is termed semi-permeable.

A saturated solution of calcium nitrate (for preparation, see exp. *a* above) is placed at the bottom of a tall glass cylinder (fig. 44), so that a column of solution about 20 cm. high is formed. 30 g. of pure phenol are shaken with 150 ml. of water in a tap funnel until equilibrium is reached, when about 20 ml. of the phenol layer (*d* = 1·05) should remain. Owing to the small difference of density between them, the phenol and aqueous layers only slowly separate, and the preparation should be left overnight, or until separation has occurred. The stem of the funnel is then projected into the cylinder, so that its end is

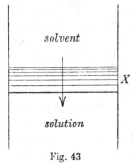

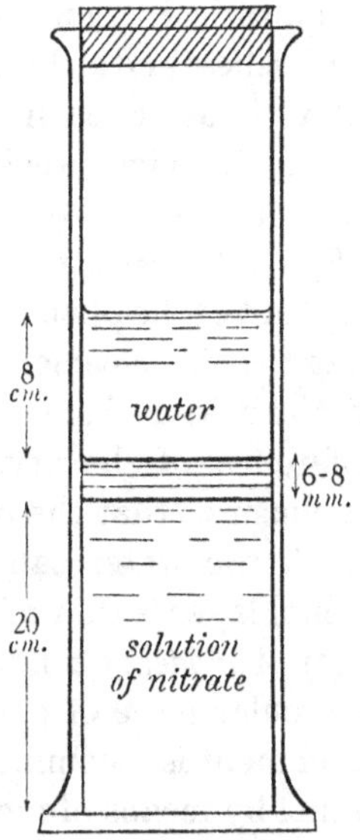

Fig. 43

Fig. 44

just above the nitrate solution and touching the side of the cylinder, and enough of the (lower) phenol layer carefully run out to form a septum 6–8 mm. thick over the solution (a layer less thick does not usually remain in position). After rejecting any remaining phenol into another vessel, the stem of the funnel is again introduced into the cylinder, and the water layer is now run out above the phenol to give a column about 8 cm. high, separated from the nitrate solution by the phenol septum. The cylinder is firmly corked to prevent evaporation, the position of the septum marked with an adhesive label, and the whole preparation set aside in a place where no great temperature changes are to be expected. The level of the phenol is observed from day to day, and is seen to rise steadily. As the solution becomes diluted by the inflow of water, the concentration gradient in the phenol becomes less steep, and the rate of rise diminishes; and although the phenol layer will ultimately reach the top of the water it may be some weeks before this final position is attained.

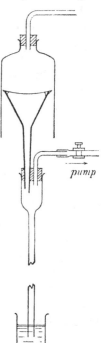

pump

Fig. 45

Since the density of the solution is 1·6, and that of the phenol 1·05, it will be clear that the displacement is due to osmosis of the water from above, and not to difference of density. It may also be emphasized that the only mechanism by which the water can continuously pass through the liquid phenol layer is that of differential solubility explained above.

(c) *A model of a natural semi-permeable membrane* (fig. 45). A circular piece of parchment paper of good quality ('sheet' parchment is recommended) is firmly attached to the rim of a funnel by means of picene wax. The stem of the funnel is provided with a double-bored rubber bung which is fitted tightly into the wide end of an air condenser; a side tube also passing through the bung is connected to a water-pump. The lower end of the condenser dips into aniline, some of which is drawn up into the tube by gentle action of the pump, and the pump connexion then closed by a screw-clip. The liquid soon falls down the tube owing to air drawn through the paper on the funnel. The liquid regains its level faster if an inverted beaker placed over the funnel is filled with the lighter gases ammonia or hydrogen, as in the familiar diffusion experiment. The paper is now thoroughly wetted, when it will be found to be gastight to air (and hydrogen) but still to permit the passage of ammonia and other very soluble gases, such as sulphur dioxide; the less soluble carbon dioxide passes more slowly. Tests with these heavier gases should be carried out by surrounding the funnel with a nearly fitting wide glass tube provided at its upper end with a bung and inlet tube for the gas.

(d) The 'chemical garden'. Fragments of ferric, cobalt, nickel and magnesium chlorides are dropped to the bottom of a beaker containing a solution of sodium silicate (10 g. of 'water-glass' to 100 ml. of water) warmed to about 25°. The thin-walled pellicles of insoluble silicates produced round the crystals expand in specific forms and colours, owing to the inward passage of water from the silicate solution to the more concentrated chloride solution within.

The experiment is most advantageously exhibited in a projection lantern, when particular attention should be paid to the narrow stream of liquid projected vertically upwards from the ruptured apices of the pellicles.

(Owing to the difficulties and laborious technique involved in experiments on the direct measurement of osmotic pressure, these are outside the scope of this book, but the student should consult standard text-books on this subject, and obtain a clear knowledge of the connexion between osmosis and osmotic pressure.)

DILUTE SOLUTIONS (*cont.*)

PART 3. THE DISTRIBUTION OF SOLUTES
BETWEEN SOLVENTS

Attention was drawn on p. 114 to the fact that Raoult's law is inapplicable to phenomena of solution in which the direct properties of solutes are concerned. We may, however, take Henry's law (p. 53) as the basis of an elementary discussion of distribution. Two solvents A and B, themselves immiscible, are *separately* shaken with a gas G at the same pressure p until each liquid is saturated. If the gas is sparingly soluble in both solvents, so that Henry's law is valid for each solution, then $p = k_1 C_1 = k_2 C_2$. On now bringing the two solutions into contact in the absence of the pure gas phase they will clearly be in equilibrium, and no change of concentration will occur. We might have achieved the same result by shaking the solvents simultaneously with the gas. If now similar experiments are made at different pressures, it will be clear that the *ratio* of the concentrations C_1/C_2 will always equal the constant k_2/k_1, although each concentration will change with the pressure. We may infer a general law of distribution $C_1/C_2 = $ constant, applicable, however, only to dilute solutions, since Henry's law applies only to such solutions (p. 53). The ratio of concentrations is called the *distribution* or *partition coefficient*.

As the solubility of a gas varies with temperature specifically for different solvents, we shall expect the coefficient also to vary with temperature.

When two solvents are shaken with a solid substance soluble in each, a similar line of reasoning will show that each becomes saturated *simultaneously*, and the distribution coefficient in this case is the ratio of the solubilities S_1/S_2. We cannot change the concentration of a solid phase as we can the pressure of a gas, but by analogy we may infer that, when the solutions are not saturated, the coefficient will still be equal to S_1/S_2 and constant, provided that even the saturated solutions are dilute. A very large amount of experimental material, of which exps. *a* and *b* are examples, confirms the validity of these theoretical predictions.

Preparation of carbon dioxide-free water and carbonate-free alkali

A supply of distilled water free from carbon dioxide will be required in the following experiments, and may be prepared most simply by drawing a vigorous current of air, by means of suction from a water-pump, first through an absorption tube filled with slightly moistened soda-lime and then, by tubes reaching to the bottom of the vessels, through one or more aspirators containing distilled water. About 3 hr. treatment is needed to prepare 10 l. of water.

To prepare sodium hydroxide solution of approximately normal concentration free from carbonate, make up first a solution of rather greater concentration, and estimate the carbonate present in a sample by the use of barium chloride as explained in exp. *c*, p. 19. Then add slightly more of a standard solution of baryta than is required to precipitate all the carbonate, protect the solution from atmospheric carbon dioxide by a soda-lime tube, and allow the precipitate to settle (but do not filter). Titrate a sample of the clear solution with standard acid, and if necessary make up with carbon dioxide-free water to the required concentration. The prepared solution should be stored in an aspirator, when the precipitate of barium carbonate will settle below the exit tube and offer no inconvenience. Alkali solutions of decinormal concentration may be prepared direct from baryta, or by treating sodium hydroxide solutions as above.

(a) The partition of acetic acid between n-butyl alcohol and water

In a bottle with a glass stopper and of about 200 ml. capacity place about 70 ml. of approximately 2*N* acetic acid, and about 50 ml. of *n*-butyl alcohol. Stopper the bottle, and well shake for at least 1 min., and then allow the two liquid layers to separate. Insert the lower stem of a 20 or 25 ml.

pipette so far through a bored cork, that when the latter rests
on the rim of the unstoppered bottle the tip of the pipette
reaches somewhat below the middle of the upper (alcohol)
layer (fig. 46). With the pipette guarded in
this way withdraw a sample of the layer, and
discharge it into a second glass-stoppered
bottle. After the addition of about an equal
volume of carbon dioxide-free water and some
phenolphthalein, titrate the sample with
standard alkali free from carbonate (a concentration about normal is suitable). During
the titration the bottle should be frequently
stoppered and well shaken, in order to transfer the acid to the water. The end-point is
indicated when a *faint* pink coloration cannot
be discharged by shaking.*

Fig. 46

To sample the water layer close the upper
stem of a second pipette (20 or 25 ml.) firmly
with the finger and thrust the tip through the
alcohol layer, of which only a drop will enter
the pipette; draw in a small quantity of the water layer and
reject again by very gentle blowing, before charging the pipette.
Titrate the sample in a flask in the ordinary way with the same
alkali as was used for the sample of the alcohol layer, again
using phenolphthalein as indicator.

25 ml. of fresh alcohol and 25 ml. of carbon dioxide-free
water (but no further acid) are then added to the liquid
remaining in the bottle, which is closed and well shaken as
before, to establish a new equilibrium with the now lesser
total quantity of acid. The procedure of sampling and titration
described above is then repeated.

The scheme below shows how further amounts of the two
solvents may be added and suitable samples withdrawn for
titration, so as to cover an adequate range of concentration.
Samples, usually 10 ml., of the weaker solutions must be

* Titration of non-aqueous solutions in all the following experiments should be carried out in this way.

titrated with decinormal alkali. For measuring the original volumes of acid and alcohol and the subsequent additional volumes of water and alcohol a graduated cylinder may be used, but the samples must of course always be taken with pipettes. As long as the same concentration of alkali is used for titrating samples of equal volume from layers in equilibrium, the partition coefficient can be found by simple division of one 'titre' by the other.

The coefficient will be that for room temperature, which should be observed and recorded. Partition coefficients do not usually change rapidly with temperature, so that a thermostat is not essential, unless it is desired to study temperatures differing much from that of the laboratory.

Initial mixture: 50 ml. n-butyl alcohol and 70 ml. $2N$ acetic acid.

Alcohol layer		Water layer		
Sample removed ml.	Fresh alcohol added ml.	Sample removed ml.	Water added ml.	Alkali for titration (both samples)
25	—	25	—	Normal
—	25	—	25	
25	—	25	—	Normal
—	25	—	70	
10	—	25 + 25*	—	Decinormal
—	0	—	50	
10	—	10	—	Decinormal

* One of the samples may be rejected without titration. In calculating the coefficient from this sampling multiply the titre of the alcohol sample by 2·5.

The coefficient at 15–20° should be $C_{water}/C_{alc.} = 0·85$. The total volume of butyl alcohol required to carry out the above series of experiments is about 100 ml. All liquors from titrations of the alcoholic samples and any other alcohol residues should be placed in residue jars, and on no account thrown into the sink. The alcohol may be easily recovered from the residues by adding common salt to nearly saturation point, and then separating the salted-out alcohol with a tap funnel. Unless the alcohol is required anhydrous for another purpose, distillation is not necessary.

(b) *The partition of ammonia between water and chloroform*

Prepare a solution of ammonia of concentration about $1 \cdot 25 N$ by diluting '880' ammonia solution in the proportion of 75 ml. for 1 l. of solution. Mix in a glass-stoppered bottle, as in exp. a, 50 ml. of this solution with an equal volume of chloroform. Stopper the bottle and well shake for at least 1 min., and then allow the layers to separate. After apparent separation has occurred gently agitate the bottle in a rotatory motion to detach lenses of the chloroform layer that usually float on the water surface. Some minute droplets of the one layer persist for a considerable time (up to $\frac{1}{2}$ hr.) in the other, but after 5 min. the error in the coefficient due to this will usually be less than 5 %. Take a 10 ml. sample of the upper (water) layer with a guarded pipette, as in exp. a, and at once deliver it into a flask containing about 100 ml. of water. Add methyl orange and titrate with standard acid ($N/2$ hydrochloric acid is suitable).

Insert a 20 ml. pipette into the lower (chloroform) layer, adopting the precaution described in exp. a. If many droplets are visible in a sample now drawn into the pipette these will have originated mainly from the walls of the bottle, which tend to remain coated with the water layer below the chloroform surface. In this case allow the sample to fall out of the pipette again, and thus flush the walls below the chloroform surface. After a short interval a second sampling should proceed without trouble. Deliver the contents of the pipette into a second glass-stoppered bottle, add water, stopper the bottle, and well shake. Titrate carefully (see footnote, p. 140), with standard acid ($N/20$ hydrochloric acid is suitable), until the pink colour of the methyl-orange indicator is permanent after shaking. When the titration is completed, decant as much of the aqueous liquid as possible, add about 10 ml. of fresh chloroform, insert a 20 ml. pipette and by its means transfer 20 ml. of the chloroform (now free from ammonia) to the partition bottle, to which 10 ml. of water is also added. The two layers being now restored to their original volumes, the bottle is stoppered, and well shaken to establish the new equilibrium with the diminished total amount of ammonia. 10 ml. of the aqueous layer and 20 ml. of the other layer are withdrawn as before, the latter being delivered into the titration bottle still containing about 10 ml. of chloroform, and some solution (neutral) remaining from the first titration. After the second titration and decantation, 20 ml. of the chloroform are as before carried back to the partition bottle. The procedure is repeated until a sufficient range of concentration has been explored (see example below). By the continued use of the same chloroform, in the way suggested, the amount required for an extensive series of experiments is limited to 60–70 ml.

Example

Original mixture: 50 ml. of approx. 1·25 N ammonia and 50 ml. of chloroform. After each sampling, numbered below, the volume of the layers was restored by the addition of 10 ml. of water, and transference of 20 ml. of chloroform from the titration bottle.

Sampling	Titrations		Coefficient = 20A/B
	(A) 10 ml. of aq. layer, with N/2 HCl	(B) 20 ml. of CHCl₃ layer with N/20 HCl	
1	27·0	22·95	23·4
2	21·0	17·15	24·4
3	16·5	13·45	24·4
4	12·95	10·45	24·8
5	10·25	8·30	24·7
6	8·1	6·5	24·9

Laboratory temperature = 19° C.

The samples were taken after only a few minutes' standing, and the slight drift in the coefficient is due to not quite complete separation.

It will be noted that the coefficient is effectively constant, although part of the ammonia in the water layer is undoubtedly chemically combined. (The value of the above coefficient will be required in exp. b, p. 147 below.)

Applications of the principle of partition

(1) *The investigation of reactions between ions and neutral molecules in solution*

A molecule M may combine in solution with an ion B (+ or −) to set up a chemical equilibrium

$$M + B\,(+\ \text{or}\ -) \rightleftharpoons M.B\,(+\ \text{or}\ -).$$

As examples we may consider the reactions to be studied in the following experiments:

$$I_2 + I^- \rightleftharpoons I_3^-, \tag{1}$$

$$4NH_3 + Cu^{++} \rightleftharpoons Cu.4NH_3^{++}. \tag{2}$$

The merely qualitative facts, that iodide solutions dissolve iodine so freely, and that the addition of ammonia to solutions of cupric salts so drastically changes the colour, leave little doubt that reactions of such a kind take place, but no direct method of chemical analysis of the solution can give information about either the extent of the combination, or the formula of the compound. The whole of the molecular iodine, free and combined in I_3^-, will be titrated by thiosulphate in (1), and the whole of the ammonia by acid in (2).

On shaking a solution containing (1) or (2) with an organic solvent, such as benzene or chloroform, etc., in which ions are insoluble, only the neutral molecules (iodine or ammonia) will become distributed. A knowledge of the partition coefficient for the chosen solvent and water can then be used to calculate the concentration of the free (uncombined) material in the chemical equilibrium, about which, as the examples will show, complete information can then be obtained. (Attention may be drawn to exp. b, p. 117, in which the cryoscopic method is employed to solve a not dissimilar problem.)

(a) *The partition of iodine between aqueous iodide and organic solvents.**

Many solvents in which iodine is soluble may be employed, e.g. carbon tetrachloride, chloroform, benzene, carbon disulphide, and ethylene dibromide, but only in the last two of these is the concentration of the iodine at equilibrium in general comparable with that in the iodide solution; of these two ethylene dibromide is selected for the experiments, owing to the danger involved in the handling of carbon disulphide.

The experiments will require about 120 ml. of a solution of potassium iodide containing 100 g. of the pure salt per litre (solution $= 0.6M$), and 40 ml. of ethylene dibromide (preferably previously washed with water). The technique is closely similar to that of exp. *b*, p. 142, which should be consulted for details, if not already carried out.

Place in the glass-stoppered bottle 50 ml. of the solution of potassium iodide, measured by means of a graduated cylinder. Add and dissolve completely by shaking 1 g. of finely powdered iodine. Add 30 ml. of ethylene dibromide, stopper the bottle and establish equilibrium by shaking, as in previous experiments. Ethylene dibromide is particularly apt to form floating lenses on the upper surface of the water, and these should be as far as possible dislodged (see exp. *b*) after the separation of the layers. Using the precautions and method of guarded pipette previously described, remove 20 ml. of the upper (aqueous) layer, dilute freely with water, and titrate with standard thiosulphate (concentration $N/10$ to $N/20$ is suitable), until the liquid is just colourless (starch may be used, if preferred).

Remove 10 ml. of the lower (dibromide) layer by the method of exp. *b*, p. 142, deliver into a second glass-stoppered bottle, add 20 ml. of water and a few crystals of potassium iodide (or 20 ml. of a 10 % solution of potassium iodide). Titrate carefully with the standard thiosulphate already used for the sample from the aqueous layer, until after shaking only a very

---

* Cf. Jakowkin, *Z. physikal. Chem.* **13**, 539 (1894).

faint violet colour is seen in the residual dibromide (starch should *not* be added). Decant most of the *aqueous* liquid from the titration bottle, add about 10 ml. of fresh dibromide, and then with a 10 ml. pipette transfer 10 ml. of the solvent to the partition bottle, to which 20 ml. of the prepared potassium iodide solution are also added (but no more iodine).

The procedure is repeated, as in exp. *b*, until a sufficient range of iodine concentration has been studied.

Example

The experiments were carried out as described above. Sodium thiosulphate $0.079N$ was used for titration.

Thiosulphate (ml.)		*Apparent* partition coefficient (r') $= A/2B$
For aqueous samples (20 ml.)	For non-aqueous samples (10 ml.)	
20·1	12·9	0·780
12·7	8·1	0·784
8·2	5·0	0·820
5·15	3·10	0·830
(*A*)	(*B*)	(*C*)

The small drift in the coefficient for a wide change of concentration at once suggests that we are concerned with a reaction of the type $nI_2 + I^- \rightleftharpoons I_{2n+1}^-$ in which $n = 1$, for then

$$\frac{[I_3^-]}{[I_2][I^-]} = K_c,$$

where $[I^-]$ is approximately constant, since it is in large excess, and for the same reason $[I_3^-]$ approximately equals the total iodine concentration in the aqueous layer. Therefore if the partition coefficient of iodine between *water* and ethylene dibromide is r ($= 0.026$) the apparent coefficient $r' \simeq r\,[I_3^-]/[I_2] = Kr[I^-] \simeq$ constant.

Evaluation of the equilibrium constant K_c *

$C_0 =$ concentration of 'fixed' iodine in the aqueous layer (g.-mol. I_2 per litre).

* Jakowkin, loc. cit.; Bray and MacKay, *J. Amer. Chem. Soc.* 32, 914 (1910); Jones and Kaplan, *J. Amer. Chem. Soc.* 50, 1845 (1928).

$C_1 =$ concentration of free iodine in the aqueous layer (g.-mol. I_2 per litre).

$C_2 =$ concentration of iodine in the dibromide layer (g.-mol. I_2 per litre).

The apparent coefficient r' is clearly given by

$$r' = (C_0 + C_1)/C_2 = C_0/C_2 + r,$$

whence

$$C_0 = (r' - r)\,C_2,$$

$$K_c = \frac{[I_3^-]}{[I_2][I^-]} = \frac{C_0}{(0{\cdot}026 C_2)[I^-]} = \frac{r'-r}{[I^-]}\frac{1}{0{\cdot}026}.$$

$r'-r$ (column C $-0{\cdot}026$)	C_2 (from titrations)	C_0 ($D \times E$)	Conc. of I^- ($0{\cdot}6 - C_0$)	K_c ($D/(G \times 0{\cdot}026)$)
0·754	0·0510	0·0385	0·5615	510
0·758	0·0320	0·0243	0·5757	502
0·794	0·01925	0·0153	0·5847	516
0·804	0·01225	0·0101	0·5899	516
(D)	(E)	(F)	(G)	

The satisfactory constancy of K_c proves that no complex ions of higher iodine content than I_3^- are formed in the solution under the conditions of the above experiments.

(b) The formula and stability of the cuprammonium kation.

A normal, i.e. semimolar, solution of cupric sulphate, and a solution of ammonia of $1{\cdot}1$–$1{\cdot}2 N$ concentration, will be required and may be prepared as follows. Weigh out into a 100 ml. graduated flask 12·50 g. of pure crystalline cupric sulphate, dissolve in water with the addition of a *few* drops of dilute sulphuric acid and make up. For the solution of ammonia, dilute 17 ml. of 0·880 solution to 250 ml. Standard acid (hydrochloric or sulphuric), about $N/2$ and $N/20$ concentrations, will be needed for titration; and about 60 ml. of chloroform for partition.

Place in a graduated cylinder of 100 ml. capacity not more than 40 ml. of the ammonia solution, add from a burette or pipette 5 ml. of the cupric solution; if necessary, stir with a glass rod until all precipitate dissolves. Then add more ammonia solution to give a total of 50 ml. of mixture. Add chloroform until the contents of the cylinder amount to 100 ml., and pour the whole of the liquids into a tap funnel, provided with a broad cork on its stem (see below). Stopper the funnel, and well shake it for at least 1 min., inverting once or twice and opening the tap to release pressure due to chloroform vapour. Allow the layers to separate (5 min.). Dislodge floating chloroform by a gentle rotatory movement of the funnel, and then with a guarded pipette (see exp. *a*, p. 140) withdraw 20 ml. of the upper blue layer, delivering into a flask containing about 100 ml. of water.

Add methyl orange, and titrate with standard acid (about $N/2$). The blue colour at first masks that of the indicator, but as the addition of acid proceeds the very faint colour of the simple cupric ion develops and in no way interferes with the normal appearance of the indicator near the end-point. A precipitate of hydroxide forms temporarily in the course of the titration. If desired a second portion of 20 ml. may be withdrawn and titrated for confirmation.

On close inspection of the (lower) chloroform layer in the funnel it will often be found that a droplet of blue aqueous solution is persisting in the cavity above the tap. If this is titrated with the chloroform layer, serious error may be introduced. To dislodge it attach a length of rubber tubing to the exit of the stem of the funnel, and blow gently while the tap is cautiously and partially opened. The entry of one or two air-bubbles will suffice; more may cause appreciable loss of ammonia from the layer. Bring an unstoppered glass bottle, containing about 50 ml. of water, below the funnel, and arrange that the cork on the stem of the latter rests on the rim of the bottle, and then discharge the whole of the chloroform layer carefully into the bottle, loss of ammonia being minimized by the cork. Add methyl orange and titrate carefully (see footnote, exp. *a*, p. 140) with standard acid (about $N/20$).

Wash away the residue of blue solution remaining in the funnel, and then pour into the latter the whole contents of the titration bottle. Place in the previously used 100 ml. cylinder 35 ml. of the original ammonia solution, add again 5 ml. of the copper solution, and after all precipitate has dissolved, add *water* to give, as before, a total of 50 ml. Run the chloroform from the tap funnel into the cylinder, to which it will be necessary to add only a few millilitres of fresh chloroform to bring the contents to 100 ml. again. Wash out the funnel, and allow it to drain for a few minutes before pouring into it the contents of the cylinder.

The procedure is repeated, but in each successive experiment a lesser volume of ammonia solution is used, as shown in the table below, but in all mixtures 5 ml. of the normal cupric solution is used, and the volume of the aqueous layer always brought finally to 50 ml. by the addition of water. Hence in all mixtures $C_{Cu} = 0.1 N = 0.05 M$.

Example

Ammonia solution in mixture ml.	Titrations		Apparent coefficient $r' = 25 \times B/C$
	20 ml. aq. layer, with $N/2$ acid	50 ml. of $CHCl_3$ layer, with $N/20$ acid	
(45)	37·25	30·0	31·0
35	29·30	22·5	32·5
25	20·50	13·0	39·5
20	18·30	10·55	43·2
(A)	(B)	(C)	(D)

In the following calculations the notation and method are parallel to those of the preceding exp. *c*, which should be consulted for details of the reasoning:

C_0 = concentration of 'fixed' ammonia (g.-mol./l.).
C_1 = concentration of free ammonia (g.-mol./l.).
C_2 = concentration of ammonia in the chloroform.

The coefficient of distribution for *water* and chloroform will be taken as $r = 24.7$, the mean of the results of exp. *b* for the range of concentration here used. The method of calculation is similar to that of exp. *c*, with the object in this case of finding directly the formula of the complex ion.

Although experimental error has brought some fluctuations into column *H*, it appears that the cuprammonium ion has the formula

Cu.4NH$_3$, but is somewhat dissociated in the concentrations of
ammonia and copper used in the above experiments.

C_2 (column C $\div$ 1000)	$r' - r$	C_6 ($E \times F$)	Ratio of conc. fixed NH$_3$ to conc. Cu^{++} ($G \div 0.05$)	C_1 ($E \times 24.7$)
0.0300	6.3	0.189	3.78	0.741
0.0225	7.8	0.178	3.56	0.515
0.0130	14.8	0.182	3.64	0.322
0.01055	18.5	0.194	3.88	0.260
(E)	(F)	(G)	(H)	

(2) *The determination of molecular state*

The fact that many substances appear to have different
molecular weights in different solvents has already been the
subject of experiments (pp. 120, 126) by the cryoscopic method.
It remains here to illustrate
the effects of such molecular
complexity on distribution.

Suppose a certain amount of
a substance X when dissolved
in each of two solvents, say
benzene (B) and water (W),
remains associated in double
molecules, but to a different
extent in each. On shaking
the two solutions together we
shall then be concerned effectively with the distribution of

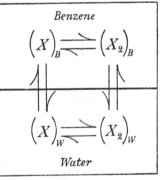

Fig. 47

two solutes, the monomeric form X and the dimeric form X_2
(fig. 47). For each there will be a characteristic partition
coefficient

$$\frac{[X]_W}{[X]_B} = r_1 \quad \text{and} \quad \frac{[X_2]_W}{[X_2]_B} = r_2. \qquad (1)$$

In each solvent there will also exist a chemical equilibrium

$$2X \rightleftharpoons X_2,$$

implying the relations, based on the law of equilibrium,

$$K_W[X]_W^2 = [X_2]_W \quad \text{and} \quad K_B[X]_B^2 = [X_2]_B. \qquad (2)$$

The relations (1) and (2) can now be combined, by substituting $[X]_B$ and $[X_2]_W$ from (2) in (1):

$$\frac{[X]_W}{\sqrt{[X_2]_B}} = \frac{r_1}{\sqrt{K_B}} = \text{const.} \quad \text{and} \quad \frac{[X]_W^2}{[X_2]_B} = \frac{r_2}{K_W} = \text{const.} \qquad (3)$$

If, as freezing-point data suggest, X_2 greatly preponderates in benzene solution, even at low concentrations, and X in water, then practical titrations of samples from the benzene and water layers effectively give $[X_2]_B$ and $[X]_W$ respectively. Consequently the simple partition relations (1) will *appear* to be replaced by (3). This argument can obviously be generalized to show that if the molecular complexity in one solvent is represented by X_m and in the second by X_n, then partition experiments will show that

$$\frac{(\text{estimated conc. in solvent 1})^n}{(\text{estimated conc. in solvent 2})^m} = \text{constant}.$$

This application of the partition principle provides a valuable means of confirming and extending other methods of discovering molecular weights in solution, particularly for solvents whose properties render them unsuited to other methods.

It should be noted that (3) is founded upon the fundamental partition principle, and in no way implies a failure of that principle.

The distribution of acetic acid between benzene and water

The apparatus and method of experiment are in every respect similar to that of exp. *a*, p. 139, on the distribution of the same acid between butyl alcohol and water.

The initial mixture in this case is prepared from about 100 ml. of approximately $3N$ acetic acid and 50 ml. of benzene. As there is a great disparity in the concentration of the acid in the two layers, as long a time as possible should be allowed for separation after shaking. In order to reduce the total amount of acid sufficiently rapidly to allow an adequate range of concentration to be examined in a reasonable time, more of the water layer must be removed than is required for actual titration. Such additional withdrawals are indicated by a *downward* arrow in the tabulated scheme.

Benzene layer (all samples titrated with $N/20$ aikali, preferably baryta)		Water layer (all samples titrated with N carbonate-free NaOH)	
Sample removed ml.	Fresh benzene added ml.	Sample etc. removed ml.	Water added ml.
↑	↑	10 + 10 ↓	
			20
		10 + 10 ↓	
In all cases 10 ml.	In all cases replace 10 ml.		20
		20 + 10 ↓	
			30
		30 + 20 ↓	
			50
↓	↓	20	

Example

20 ml. of the baryta solution used for the benzene samples required 10·9 ml. of $N/10$ hydrochloric acid for neutralization. The concentration of the baryta was therefore 0·0545N.

It will be seen that the drift in column 2 is in the opposite direction to and is much less in magnitude than that of column 1. Thus we should expect to find strict constancy in a partition expression which allowed for the presence of a small but not negligible proportion of monomeric molecules in the benzene solution.

This result shows some apparent contrast with the results of freezing-point experiments, which indicate a quite negligible fraction of mono-

Titrations			C_B (g.-mol. per litre)	C_W (g.-mol. per litre)
10 ml. of benzene layer with baryta		Samples of water layer (N NaOH)		
1	35·4	10 ml. = 30·6	0·193	3·06
2	23·8	10 ml. = 24·2	0·130	2·42
3	16·2	10 ml. = 19·5	0·0885	1·95
4	8·6	20 ml. = 27·5	0·0471	1·35
5	2·8	20 ml. = 14·2	0·01525	0·71

C_W/C_B	$C_W/\sqrt{C_B}$
15·9	6·96
18·6	6·70
22·0	6·65
28·6	6·20
46·6	5·75
(1)	(2)

meric molecules even at lower concentrations than those treated above, but it must be remembered that in the partition experiments the benzene is necessarily saturated with water, and although the amount of water so dissolved (0·211 %) is small, it nevertheless profoundly modifies the solvent properties of the hydrocarbon. This important consideration is often lost sight of in comparing results of the two methods.

Chapter V

THERMOCHEMISTRY

The heat change ΔH in a reaction may be determined calorimetrically in two principal ways: (i) when the reaction takes place, or can be made, with the aid of catalysts, to take place, sufficiently rapidly in (aqueous) solution, the heat change may be calculated from the rise of temperature, suitable precautions being taken to counteract the effect of heat losses; (ii) when combustible substances are concerned, heats of reaction may be determined as the difference of the heats of complete combustion of the reactants and resultants. The heat of the reaction

$$H^+(aq.) + OH^-(aq.) = H_2O \text{ (liq.)} \tag{1}$$

may be obtained by mixing in a calorimeter dilute solutions of a strong base (KOH, NaOH) and a strong acid (HCl, HNO$_3$). If the reacting solutions are sufficiently dilute, the complete equation $M^+ + OH^- + H^+ + X^- = H_2O + M^+ + X^-$ clearly reduces to (1) (exp. 2, a). The law of constant heat summation is frequently invoked in cases where a direct calorimetric measurement would be difficult or impracticable; thus the heat of the hydration

$$Na_2CO_3 \text{ (solid)} + 10H_2O \text{ (liq.)} = Na_2CO_3 . 10H_2O \text{ (solid)}$$

is found by determining in separate experiments the heats of solution of anhydrous sodium carbonate and of the decahydrate (exp. 3). The law is also assumed in operating method (ii) above. For example, the heat change ΔH_0 in

$$C \text{ (graphite)} + 2S \text{ (rhombic)} = CS_2 \text{ (liq.)}, \quad (\Delta H_0)$$

is found by combining separately determined heats of combustion as follows:

$$C \text{ (graphite)} + O_2 \text{ (gas)} = CO_2 \text{ (gas)}, \quad (\Delta H_1)$$
$$S \text{ (rhombic)} + O_2 \text{ (gas)} = SO_2 \text{ (gas)}, \quad (\Delta H_2)$$
$$CS_2 \text{ (liq.)} + 3O_2 \text{ (gas)} = CO_2 \text{ (gas)} + 2SO_2 \text{ (gas)}, \quad (\Delta H_3)$$
$$\Delta H_0 = \Delta H_1 + 2\Delta H_2 - \Delta H_3.$$

Signs, notation and units

The sign of a heat change, like that of other energy changes, is decided by reference to the loss or gain *by the system*, and not by the corresponding changes in the surroundings (see also p. 65). In exothermic and endothermic reactions the heat changes are respectively negative $(-\Delta H)$ and positive $(+\Delta H)$. For the case of CS_2 above

$$\Delta H_3 = -262 \cdot 2 \text{ kcal.,}$$
$$\Delta H_2 = - 70 \cdot 9 \text{ kcal.,}$$
$$\Delta H_1 = - 94 \cdot 3 \text{ kcal.,}$$

whence $\Delta H_0 = -236 \cdot 1 + 262 \cdot 2 = +26 \cdot 1$ kcal., and the direct reaction between graphite and solid sulphur would be endothermic.

No precise meaning can be attached to the value of a heat change unless the equation of the reaction and the physical and chemical states of the reactants and resultants are given, as in the above examples. In equations of combustion the oxygen is to be expressed in molecules, e.g. $1\frac{1}{2}O_2$, not $3O$. The suffix (gas) is taken to imply that the pressure is 1 atm. In the case of aqueous solutions the suffix (aq.) or (∞) means that the solution is so dilute that no measurable heat change results from further dilution; when this is not the case the composition of the solution must be stated, preferably in terms of gram-molecules of the solvent, e.g. H_2SO_4 (55·5 aq.) or $H_2SO_{4(55\cdot5)}$ means a solution of 98 g. of pure sulphuric acid in 55·5 g.-mol. $= 1000$ g., of water, i.e. an approx. 2-normal solution.

Heat changes are commonly expressed in one of two units: (*a*) the *kilo-calorie* (kcal.), defined as the amount of heat required to raise 1 kg. of water from 15 to 16° C.; (*b*) the *kilo-joule* (kj.) $= 10^{10}$ ergs $= 0\cdot2391$ kcal. The latter unit is convenient when heat energy has to be compared with or related to other forms of energy.

In principle, the heat change of a reaction is not completely defined unless the external physical conditions—constant pressure or constant volume—and the temperature at which the reaction is carried out are stated. The notation ΔH or ΔU is used for reactions at constant pressure (1 atm.) or constant volume respectively: e.g.

$$CH_3OH \text{ (liq.)} + 1\tfrac{1}{2}O_2 \text{ (gas)} = CO_2 \text{ (gas)} + 2H_2O \text{ (liq.)}.$$
$$\Delta H_{298} = -173 \cdot 63 \text{ kcal.}$$
$$\Delta H_{298} = -173 \cdot 33 \text{ kcal.}$$

However, only in exceptional cases do ΔH and ΔU differ by more than a small fraction of the value of either; the influence of small changes of temperature on ΔH or ΔU may in general be neglected, but may when necessary be calculated from the relation

$$\frac{d(\Delta H)}{dT} = \Sigma c_p, \quad \frac{d(\Delta U)}{dT} = \Sigma c_v,$$

where c is the molecular heat of a reactant or resultant, and $\Sigma c = \Sigma c_{\text{resultants}} - \Sigma c_{\text{reactants}}$.

It is necessary to emphasize that the possibility and direction of physical and chemical reactions are not in principle determined (as was originally asserted by Thomsen and by Berthelot*) by the heat change ΔH, but by the sign and magnitude of the free-energy change ΔG. Reactions can proceed spontaneously only in the direction in which ΔG is negative, i.e. when $d(\Delta G) < 0$; and the tendency to reaction is measured by the absolute magnitude of ΔG. Free-energy change ΔG and heat change ΔH are connected by the thermodynamic relation

$$\Delta G = \Delta H - T\Delta S, \tag{2}$$

where ΔS is the entropy change. Since $[\partial(\Delta G)/\partial T]_p = -\Delta S$, (2) may be rewritten

$$\Delta G - \Delta H = T\left[\frac{\partial(\Delta G)}{\partial T}\right]_p, \tag{3}$$

or again, in the more compact and mathematically convenient form,

$$\frac{\partial(\Delta G/T)}{\partial T} = -\frac{\Delta H}{T^2}. \tag{4}$$

* *Essai de mécanique chimique*, Paris, 1878.

According to (2) ΔG may be negative, and a tendency to react exist even when ΔH is positive (endothermic reaction), if ΔS is large and positive. The simplest cases of this kind are the physical processes of melting, evaporation and dissolution, where the large gain of entropy on formation of the more disordered state (liquid or vapour) offsets the positive sign of ΔH. In many organic reactions ΔH, although negative, is small, and the tendency to reaction again depends mainly on the second member of (2). It may, however, be said that for the majority of chemical reactions a negative sign and large absolute magnitude of ΔG are secured by a large negative ΔH, and to this extent the original theorem of Berthelot and Thomsen is true.

When ΔH is known as a function of T, it is possible in principle to calculate ΔG from (4), except for an integration constant, which can also be determined if ΔG is known at one temperature. The limiting condition $d(\Delta G) = 0$ is that of no change, i.e. is the condition of equilibrium. In systems where the individual contribution of the constituents (A, B, etc.) is given by $d(\Delta G_A) = RT d \ln p_A$, etc., this condition leads at once to the general law of chemical equilibrium

$$\Delta G_T = -RT \ln K_p, \qquad (5)$$

in which K_p is the equilibrium constant expressed as the quotient of products of partial pressures, e.g.

$$K_p = \frac{p_R^r p_S^s, \text{ etc.}}{p_A^a p_B^b, \text{ etc.}}.$$

Hence, from what has been said above, if K_p is found for one temperature, ΔG, and therefore, from (5), K_p can be calculated for all temperatures. We obtain also, by substituting (5) in (4), the well-known relation

$$\frac{\partial \ln K_p}{\partial T} = \frac{\Delta H}{RT^2}, \qquad (6)$$

or, assuming $\Delta H \neq f(T)$,

$$\ln \left\{ \frac{(K_p)_{T_1}}{(K_p)_{T_2}} \right\} = \frac{\Delta H}{R} \left\{ \frac{1}{T_2} - \frac{1}{T_1} \right\}. \qquad (7)$$

From (7), ΔH may be calculated, as an average (constant) value for the temperature range concerned, from two values of K_p, at T_1 and T_2.

(For the electrical method of finding ΔG directly, see Chapter VI, p. 197.)

The experiments to follow will be confined to simple examples of reaction in aqueous solution; for the technique of determining heats of combustion in the 'bomb' calorimeter the reader is referred to larger works.

If a rise of temperature $\Delta t°$ (corrected for heat losses) is observed in a solution weighing W g. contained in a calorimeter of water-equivalent w g., the corresponding heat change q is given by $q = (W \times c + w)\Delta t$, where c is the specific heat of the solution. The specific heats of the majority of aqueous solutions diminish, while the specific gravities rise, with increase in concentration. In general, for dilute solutions (usually to about molar concentration), these two effects offset each other to such an extent, that to the accuracy attainable with simple apparatus, i.e. 1–2 %, the water-equivalent of a solution may be taken as equal to the water contained in it (see table 17, for solutions of sulphuric acid).

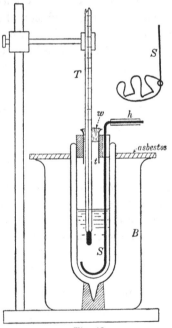

A calorimeter suitable for general semi-quantitative experiments with solutions may be simply constructed by loosely packing the space between two large boiling tubes or two beakers with cotton-wool. The modern vacuum flask has the merit of very good heat insulation, but the narrowed neck of the ordinary

Fig. 48

Thermos type prevents the use of an efficient stirrer, and, if mixing is accomplished by shaking, the water-equivalent becomes large and quite indefinite in value. An unsilvered, straight-sided Dewar vessel (fig. 48) avoids this serious disadvantage without notable loss of heat insulation, and in addition its transparency is invaluable in experiments on heats of solution, etc. The vessel should be closed with a cork or rubber bung, grooved for the stirrer S, and carrying a short wide tube t, through which passes the thermometer (graduated at least in 0·1°), which is supported externally by a retort clamp. The tube t should be stoppered with a plug of cotton-wool, and serves in all experiments as inlet for the introduction of liquids or solids into the calorimeter. The stirrer S should be as efficient as possible, and may be formed of copper or silver wire bent as shown in fig. 48, S. It should finally be given a hemispherical form so that it fits the rounded base of the calorimeter tube. A sleeve of stout rubber, or ebonite, tubing h serves as a heat-insulating handle. An empty glass beaker B placed as a shield round the calorimeter usually improves the regularity of the temperature observations. When it is necessary to have water warmed considerably above room temperature in the calorimeter (as in exp. a, p. 170), the shield should be filled also with warm water.

(1) *The determination of the calorimeter constant*

Before the calorimeter can be used to determine the heat change of a given reaction it must be calibrated to establish the connexion between temperature change of its aqueous contents and the córresponding amount of heat. The heat concerned in effecting a rise (or fall) of 1° may be termed the *calorimeter constant*. Owing to the poor thermal conductivity of glass this constant varies in practice with the area of glass in contact with the liquid contents, i.e. with the volume of the contents. It is therefore essential to calibrate with that volume of water in the calorimeter that is to be used in subsequent experiments.

For the calibration it is necessary to impart to the calorimeter and its contents a known amount of heat. If the necessary electrical equipment is available, this can be very accurately achieved by passing a measured current I through a wire of known resistance R immersed in the water of the calorimeter, when the heat liberated is given by $q = RI^2\tau$, τ being the time for which the current is passed. Another very simple and convenient method, described in detail below, utilizes the heat of solution of sulphuric acid, which has been very exactly determined by several observers, notably by Pickering.*

Place in the calorimeter with the aid of pipettes or a burette a measured amount of distilled water, which should not be less than 50 ml. Set the cotton-wool stopper w in the broad tube t, and make any necessary adjustments to ensure that the stirrer can be freely operated. Pour into a small glass cylinder a supply of concentrated sulphuric acid ($d = 1\cdot84(3)$, 98·5 % H_2SO_4) from a store that has not been unduly exposed to atmospheric moisture (do not take the supply from a small reagent bottle but preferably from a Winchester quart), and insert into the acid a graduated dropping tube. Estimate from table 17 the approximate volume of acid required to give the desired rise of temperature (3–5°), remembering that the table gives amounts for addition to 100 ml. of water.

Stir the water in the calorimeter *slowly* and regularly by moving the stirrer up and down through the whole volume of the liquid about once in 2 sec.; note and at once plot in a chart on squared paper the temperature at intervals (1 min.), until it is apparent that any rise or fall is occurring at a constant rate (fig. 49). At the end of an interval quickly open the calorimeter by removing the cotton-wool stopper w, insert the dropping tube, and deliver acid into the water as nearly as possible 30 sec. after the last temperature reading. After closing the calorimeter with the stopper resume the stirring and temperature observation as before, until the rate of fall has become constant (10 min.).

Pour the contents of the calorimeter into a flask, and cool

* *J. Chem. Soc.* 57, 110 (1890).

to room temperature, before titrating a sample (25 ml.) with a standard (normal) solution of alkali. Calculate the normality of the acid solution. Project backwards to the time of mixing the linear portion *CD* (fig. 49) of the temperature chart, and

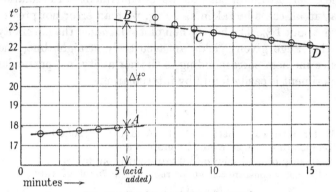

Fig. 49. Graphical calorimetric record.

so obtain the corrected temperature rise Δt. During the interval *BC* the glass of the calorimeter and, thermometer is (slowly) coming into temperature equilibrium with the aqueous contents: *AB* gives the rise which would have occurred if heat liberation and transfer had been instantaneous at the moment of mixing.

Table 17. *Heat liberated on adding concentrated sulphuric acid ($d = 1·84$, 98·5 % H_2SO_4) to 100 ml. of water*

Acid added ml. ($d = 1·84$)	Normality of final solution	Heat liberated kcal.	Specific heat of solution	Weight of solution g.	True water equivalent
3·0	1·100	0·946	0·959	105·5	101·2
2·5	0·918	0·784	0·966	104·6	101·0
2·3	0·842	0·718	0·970	104·2	101·0
1·5	0·552	0·473	0·981	102·7	100·8
0·75	0·277	0·242	0·991	101·3	100·6
0·6	0·217	0·191	0·995	101·1	100·6

The data in table 17, which are calculated from the work of Pickering,* show in column 3 the heat liberated when ordinary concentrated sulphuric acid ($98 \cdot 5 \% \ H_2SO_4$) is added to 100 ml. of water, to yield a solution of the normality shown in col. 2. The other data show that the water equivalent of the solution may in all cases be taken as 100 g. with an error not greater than about 1 %. The heat, when plotted against the normality in the range given, yields an accurately straight line, and hence the heat corresponding to the normality in a particular experiment is easily found by interpolation.

Example

Water in the calorimeter = 75 ml.   Acid added, about 1 ml.
Temperature rise, corrected by graph = $3 \cdot 77°$.
Normality of final solution = $0 \cdot 490$; heat (from table) = $0 \cdot 422$ kcal.
Calorimeter constant = heat for rise of 1° in 75 ml. of water

$$= 0 \cdot 422 \times \frac{75}{100} \times \frac{1}{3 \cdot 77} = 0 \cdot 084(0) \text{ kcal./degree.}$$

In experiments in which concentrated solutions are formed or used, or for other reasons, it may be necessary to know the water equivalent of the calorimeter alone. This may at once be calculated from the above data as follows:

Heat to raise 75 ml. (= 75 g.) of water $3 \cdot 77° = \dfrac{75 \times 3 \cdot 77}{1000} = 0 \cdot 283$ kcal.

Actual heat liberated $= 0 \cdot 422 \times \dfrac{75}{100} = 0 \cdot 317$ kcal.

Water equivalent of the calorimeter, etc., alone

$$= (0 \cdot 317 - 0 \cdot 283) \frac{1000}{3 \cdot 77} = 9 \cdot 00 \text{ g.}$$

(2) *The heat of neutralization*

To attain with simple apparatus a fair accuracy in a calorimetric measurement of a heat of reaction in solution between two substances A and B, it is advisable to arrange if possible that a dilute solution of one reagent (say A) is first placed in the calorimeter, to which a small volume of a concentrated solution of B is added when temperature conditions in the

* *J. Chem. Soc.* **57**, 110 (1890).

calorimeter are steady. The heat of dilution of the solution of B is then determined in a second parallel experiment with water in the calorimeter. Although this procedure involves two experiments, it has the advantage of needing only one piece of heat-insulated apparatus, and one accurate thermometer: it allows mixing to be accomplished almost instantaneously, and uncertainties introduced by the transference of relatively large volumes of liquid from one heat-insulated enclosure to another are eliminated.

(a) *The heat of neutralization of a monobasic acid (nitric acid) by sodium hydroxide.*

(i) Put into a graduated cylinder 65 ml. of ordinary pure concentrated nitric acid ($d = 1\cdot42$, $HNO_3 = 70$ %) and dilute to 100 ml. with water. Well mix and cool to room temperature before using in the experiments. The $10N$ solution so prepared has $d = 1\cdot31$, and contains 50 % of water.

Mix in a calorimeter (of known constant, see exp. 1) normal sodium hydroxide and water in the proportion of 25 ml. of the alkali to 50 ml. of water. Make temperature observations and plot a chart as in exp. 1. At a known time add from a dropping tube or small burette 2·5 ml. of the prepared nitric acid solution for every 25 ml. of normal soda solution placed in the calorimeter. Continue the observations of temperature, and plot a chart, also as in exp. 1.

When the temperature observations are completed add to the solution in the calorimeter a few drops of methyl orange. If the liquid proves to be alkaline, titrate it in the calorimeter with standard ($N/10$) acid: if acid, then it will be known that 25 ml. of normal alkali have been neutralized by an equivalent of acid, and no titration is necessary.

(ii) Determine the corrected rise of temperature when the volume of acid used in (i) is added to a volume of distilled water equal to the total volume of liquid used in (i). The rise of temperature will not exceed 0·5°.

Example

(i) Content of calorimeter = 50 ml. of water + 25 ml. of N NaOH ($d = 1.04$).

Total water = 75 g.

Nitric acid added = 2.5 ml., containing 50 % water.

Water = 1.25 g.

Total final content of water = 76.2 g.

Constant of calorimeter (for 75 g.) = 0.0840 kcal./degree: for 76.2 g. taken as = 0.0857.

Room temperature during experiment = 19.5–19.8°.

Temperature in calorimeter at time of mixing = 19.10°.

Corrected temperature rise (Δt) = 4.18°.

Final solution reacted alkaline, and required 16.0 ml. $N/10$ HCl.

$$\text{G.-mol. of base and acid neutralized} = \frac{25 - 1.6}{1000} = 0.0234.$$

(ii) *Heat of dilution of acid.* Temperature rise on addition of 2.5 ml. of acid to 74 ml. of water = 0.41°.

Temperature rise on neutralization, corrected for heat of dilution

$$= 4.18 - 0.41 = 3.77°.$$

Heat of neutralization per g.-mol. of acid and base

$$= \frac{0.0857 \times 3.77}{0.0234} = 13.80 \text{ kcal.}$$

$$(\Delta H = -13.80 \text{ kcal.})$$

(Additional experiments may be carried out by the above method, and a general survey of the subject made. Suitable acids which may be used include sulphuric, hydrochloric, perchloric and acetic: ammonia may be substituted for the strong bases: Table 18.)

(b) *The heat of neutralization of polybasic acids* (*phosphoric acid*)

The following solutions will be required, and should be prepared some time beforehand, so that they may have assumed the temperature of the laboratory when the experiments are begun:

(1) Normal sodium hydroxide and normal hydrochloric acid.

(2) $M/2.5 \, Na_2HPO_4$: dissolve 28.7 g. of fresh (un-effloresced) crystals ($12H_2O$) for 200 ml. of solution. This solution is nearly saturated.

Table 18. *Heats of neutralization of acids by sodium hydroxide (kcal.)*

NaOH.nH$_2$O + HX.nH$_2$O: approx. normality 55·5/n.

n	HCl (20°)*	HNO$_3$ (20°)*	$\frac{1}{2}$H$_2$SO$_4$‡	C$_2$H$_4$O$_2$ (20°)†	HClO$_4$§
25	14·228	14·012	15·850	—	—
50	14·009	13·892	15·685	13·375	—
100	13·895	13·837	15·600	13·459	—
200	13·825	13·790	15·425	13·510	14·080
400	13·761	13·756	—	13·539	—
∞ *	13·660	13·705	—	13·650	—

(3) MKH$_2$PO$_4$: dissolve 27·2 g. of the salt for 200 ml. of solution.

(4) M/2H$_3$PO$_4$: mix equal volumes (50 ml.) of normal HCl and the solution of KH$_2$PO$_4$ (3).

A simple calorimetric apparatus, affording results of sufficient accuracy, is shown in fig. 50. The inner cup V is a boiling tube shortened by cutting off and rounding in the blowpipe; it should have a capacity of 15–20 ml. and be of such length that it can lie horizontally on the base of the conical flask F (200–250 ml.). The wooden box W serves as a thermally insulated enclosure, and should be provided with a lid cut in two pieces (conveniently from asbestos board) to fit the neck of the flask closely.

Assemble the apparatus as shown in fig. 50, the cup V being held in position partly by the thermometer and partly by a thin wire or thread clipped in the smaller cork bung b. Withdraw the large bung B, with the thermometer and cup in

* Richards and Rowe, *J. Amer. C.S.* **44**, 684 (1922).
† Richards and Mair, *ibid*, **51**, 737, (1929).
‡ Mathews and Germann, *J. Physical Chem.* **15**, 73 (1911).
§ Thomsen, *Thermochem. Untersuchung*, 1882.

position, support in a retort clamp, and place in V 15 ml. of normal hydrochloric acid. Into the flask put 75 ml. of normal sodium hydroxide, and then carefully lower the thermometer and cup into the flask, set the bung B firmly in position, and if necessary slightly adjust the cup so that it rests on the bottom of the flask. Place the whole apparatus in the box W,

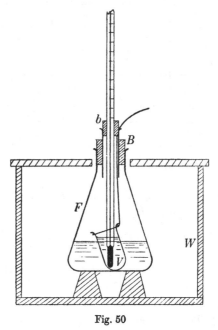

Fig. 50

and stir the contents of the flask by occasional *gentle* rotation. When the temperature has become practically constant, it is recorded and the time noted. Lift the thermometer by removing the *smaller* bung b, and thus allow the cup to fall into a horizontal position. Replace the thermometer, and well mix the contents of the cup with the alkali by removing the apparatus for a short time from the box and rotating in an inclined position. Replace in the box as soon as possible, stir by rotation as before, and plot a temperature chart as in the previous experiments (10 min.).

Repeat the procedure, placing successively in the cup 15 ml. of solutions 2, 3 and 4, using always 75 ml. of normal sodium hydroxide in the flask. Estimate the corrected rise of temperature in each experiment by projection of the chart (as in exp. 1 above).

Example

75 ml. normal sodium hydroxide in the flask and 15 ml. of solution in the cup.

Acid neutralized	Concentration of acid (C)	Δt°	$\Delta t/C$	Heat evolved per g.-mol. of acid (arbitrary units — HCl = 100)
HCl	1·0	2·07	2·07	100
$HPO_4^=$	0·4	0·47	1·18	57
$H_2PO_4^-$	1·0	2·96	2·96	143
H_3PO_4	0·5	2·56	5·12	248

The large excess of alkali used ensures that neutralization, even of the last stage ($HPO_4^=$) approaches completion, in spite of hydrolysis. By subtracting the values in column 5 in succession the heat for the separate stages is obtained, as in the table below: the second column shows the actual heat calculated on the assumption that the heat liberated in the experiment with hydrochloric acid is 13·8 kcal./g.-mol.

Stage of neutralization	Heat (HCl = 100)	Heat kcal
$HPO_4^= \rightarrow PO_4^{\equiv}$	57	7·9
$H_2PO_4^- \rightarrow HPO_4^=$	86	12·0
$H_3PO_4 \rightarrow H_2PO_4^-$	105	14·6

The values in the last column are only approximate, as no allowance has been made for heats of dilution; the experimental method is hardly accurate enough to allow these to be evaluated. It will be seen that the heat of the first stage of neutralization differs little from the values for strong monobasic acids, but an increasingly smaller value is given for the succeeding stages. Such a behaviour is characteristic of polybasic acids.

The following results are given by Thomsen:

			kcal.
H_3PO_4	$\rightarrow$	$H_2PO_4^-$	14·80
$H_2PO_4^-$	$\rightarrow$	$HPO_4^=$	12·3
$HPO_4^=$	$\rightarrow$	$PO_4^{\equiv}$	8·2

(c) *The mutual neutralization of very weak acids and bases*

When very weak acids and bases are brought together in solution, the extent of the neutralization, which may be very far from complete, may be conveniently estimated by a thermal method. If the degree of neutralization is n, the degree of hydrolysis x of the corresponding salt is clearly $1 - n$, and from this value the dissociation constant of a weak base or acid may be calculated, by using the formula for degree of hydrolysis appropriate to the conditions, viz.

$$\frac{x}{1-x} = \sqrt{\frac{K_w}{K_a K_b}}.$$

$K_w =$ the ionic product for water $= 10^{-14}$ at room temperature; K_a and K_b are the dissociation constants for the acid and base respectively. The degree of hydrolysis x is nearly independent of dilution.

The interaction of aniline and acetic acid.

(i) Determine the heat (ΔH_1) of the reaction

$$C_6H_5NH_2 \,(\text{liq.}) + H^+ \,(\text{aq.}) = C_6H_5NH_3^+ \,(\text{aq.}).$$

Place in the calorimeter (pattern of fig. 48) 75 ml. of normal hydrochloric acid. Using the methods already described in detail in the preceding experiments, find the corrected rise of temperature when 5 ml. ($= 5\cdot1$ g.) of pure dry aniline are dissolved in this acid. Calculate from the known constant of the calorimeter (see exp. 1), and assuming that the water equivalent of its contents $= 75$ g., the corresponding heat change (q). Then $\Delta H_1 = \dfrac{93}{5\cdot1} q$. (Unless the temperature of the aniline at mixing differs appreciably from that of the acid, its specific heat

(0·50) will not be required.) Since a 50 % excess of strong acid is used in the neutralization, the hydrolysis of the salt formed (hydrochloride) is negligible.

(ii) Place in the calorimeter 55 ml. of normal acetic acid (1 equiv.) and 20 ml. of water. Determine as in (i) the heat change when 5 ml. of aniline (1 equiv.) are dissolved in this acid (the liquid should be rather vigorously stirred for 30 sec. after the base has been added). Hence calculate the heat change when 1 g.-mol. (93 g.) of aniline is dissolved by 1 g.-mol. of acetic acid in approximately 0·7 M concentration.

Assuming that the heat of ionization of acetic acid in about normal concentration is 0·30 kcal. (see table 18, p. 165), and that the acid is negligibly ionized in this concentration, calculate the heat change ΔH_2 of the reaction

$$C_6H_5NH_2 \text{(liq.)} + H (H_3C_2O_2) \text{(aq.)}$$
$$= C_6H_5NH_3^+ \text{(aq.)} + H_3C_2O_2^- \text{(aq.)}.$$

(iii) Using a more concentrated solution of acetic acid (3-normal) make further experiments to determine the heat change when aniline is dissolved in two and three equivalents of acetic acid.

Example

Exp. i. Corrected $\Delta t = 4·70°$.

$$\text{Calorimeter constant} = 0·0822 \text{ kcal./degree.}$$
$$q = 0·387 \text{ kcal.}$$
$$\Delta H_1 = 7·06 \text{ kcal.}$$
$$\Delta H_2 = 7·06 - 0·30$$
$$= 6·76 \text{ kcal.}$$

Exp. ii. Heat liberated when 1 g.-mol. of base was mixed with

1 g.-mol. of acetic acid $(0·7 N) = 3·26$ kcal.

Degree of neutralization $(n) = \dfrac{3·26}{6·76} = 0·48.$

Degree of hydrolysis of aniline acetate $(x) = 1 - n = 0·52.$

K_a for acetic acid $= 1·8 \times 10^{-5}.$

K_b for aniline $= 0·47 \times 10^{-9}.$

In the derivation of the hydrolysis formula it is assumed that K_b is defined as

$$K_b = \frac{[B.H^+].[OH^-]}{[B]} \qquad (B = \text{base}).$$

It will be noted that for a base of the NH_3 type this definition does not *necessarily* imply the existence of any undissociated hydroxide, such as NH_4OH, but only of the equilibrium $NH_3 + H^+ \rightleftharpoons NH_4^+$. For a discussion of recent views of acids and bases, see Bell, *The Use of the Terms 'Acid' and 'Base', Quart. Rev.* **1**, 113 (1947): *Acids and Bases*, (Methuen) 1952.

Exp. iii. Owing to the suppression of hydrolysis by excess of acid, the heat liberated will increase with the amount of acid used; with a sufficient excess of acid the value will approach ΔH_2.

(3) Heat of solution*

(a) *The heat of solution of* Na_2CO_3

When an anhydrous salt is added in bulk to cold water a considerable part of the solid usually sets at once to a very hard mass of hydrate, which then dissolves very slowly. To overcome this difficulty with certainty the temperature of the water in the calorimeter must be initially above the transition temperature hydrate-anhydrous salt (in this case about 35°). To minimize heat losses under these conditions the calorimeter is surrounded with a bath of water kept at about the same temperature.

Gently ignite 8·5 g. of pure sodium bicarbonate in a small evaporating dish: allow the resulting anhydrous carbonate (5·4 g. = about $M/20$) to cool to room temperature in a desiccator, then transfer to a stoppered weighing bottle and weigh to 1 cg. This method of preparation gives a very finely divided specimen of sodium carbonate suitable for the calorimetric experiment.

Fill the shield of the calorimeter (fig. 48) with water, place a burner beneath, and heat, with stirring, to about 45°, when the burner is turned off. Put in the calorimeter 75 ml. of water previously heated to about 40°. (This volume of

* See notes, pp. 66, 67.

water is suitable for the quantity of carbonate prepared as described above.) Stir the water in the calorimeter, and begin temperature readings, if necessary gently reheating the water in the shield so that when steady conditions are reached the temperature of the shield is about 5° above the calorimeter temperature, which should be slowly rising, or stationary. At a known time note the room temperature and then rapidly add the specimen of carbonate to the calorimeter in one portion through the inlet tube w (fig. 48). Close the calorimeter, and stir vigorously for 30 sec., at the end of which the salt should have dissolved (do not attempt to dislodge any small amount of salt adhering to the upper part of the thermometer, or to the inlet tube). Continue stirring at the normal rate, plot a temperature chart, and so obtain the corrected temperature rise, as in previous experiments.

When the temperature record is completed, carefully remove the thermometer and inlet tube, wash down any adhering carbonate into a beaker with distilled water, and estimate its weight by titration with $N/10$ acid (methyl orange).

Example

Weight of carbonate prepared = 5·23 g.
   Bath temperature = 45–44°.
   Room temperature at mixing = 23·3°.

Temperature record. Before mixing, constant at 39·02°. Carbonate dissolved completely 25 sec. after addition. Temperature readings, resumed (at 1 min. intervals) 1 min. after first opening calorimeter, were as follows:

42·88, 42·80, 42·70, 42·65, 42·60, 42·55, 42·50, 42·45.

$$\Delta t = 3·80°.$$

Residual undissolved carbonate = 0·04 g.
Net weight dissolved in 75 ml. of water = 5·19 g.
Specific heat of solid $Na_2CO_3 = 0·25$. Heat required to raise 5·19 g. from room temperature 23·3° to mixing temperature 39·02°

$$= \frac{15·7 \times 5·19 \times 0·25}{1000} = 0·0204 \text{ kcal.}$$

Calorimeter constant (from previous experiment) = 0·084 kcal./degree.

Total heat liberated = (0·084 × 3·80) + 0·0204 = 0·339 kcal.

Heat per g.-mol. carbonate $= \dfrac{0.339 \times 106}{5.19} = 6.90$ kcal.

$$\text{Na}_2\text{CO}_3 \text{ (solid)} + 83\text{H}_2\text{O} = \text{Na}_2\text{CO}_{3(83)}, \quad \varDelta H_1 = -6.90 \text{ kcal.} \quad (1)$$

(b) The heat of solution of $\text{Na}_2\text{CO}_3 . 10\text{H}_2\text{O}$

Select from a sample of soda crystals 14–15 g. ($M/20 = 14.3$) of material showing little or no efflorescence. Crush to a *coarse* powder (loss of water will result from too long or too vigorous grinding), place in a weighing bottle which is at once stoppered and weighed. Immerse the bottle in a beaker of cold water provided with a common thermometer, in order that the solid may be at a definite temperature at dissolution. For the above weight of hydrate, put into the calorimeter (pattern of fig. 48) 65 ml. of distilled water, warmed about 6° above room temperature. After recording on a temperature chart the initial cooling line, note the temperature of the water surrounding the specimen of hydrate, and then quickly introduce the latter through the inlet tube w (fig. 48) into the calorimeter in one portion. Stir vigorously until all is dissolved (1 min.), then adopt the normal rate of stirring and plot the *heating* line.

Example

Weight of hydrate $= 14.69$ g.
Temperature of hydrate at time of mixing $= 20.3°$.
Temperature in calorimeter at time of mixing $= 25.3°$.
Specific heat of hydrate $= 0.50$.
Heat required to raise hydrate to calorimeter temperature

$$= \frac{14.7 \times 5.0 \times 0.5}{1000} = 0.037 \text{ kcal.}$$

Corrected $\varDelta t = -9.76°$.

Final water content of calorimeter:
Water originally added $= 65$ g.
Water content of hydrate $= 14.7 \times \dfrac{180}{286} = 9.25$ g.

$$\text{Total} = 74.25 \text{ g.}$$

Calorimeter constant (for 75 g.) $= 0.084$ kcal./degree, taken as 0.0835 in this experiment.

Net heat absorbed in forming the solution
$$= (0{\cdot}0835 \times 9{\cdot}76) - 0{\cdot}037 \text{ kcal.} = 0{\cdot}778 \text{ kcal.}$$

Heat per g.-mol. of hydrate ($M = 286$) = 15·15 kcal.

$$\mathrm{Na_2CO_3 . 10H_2O \ (solid)} + 70\mathrm{H_2O} = \mathrm{Na_2CO_{3(80)}}, \quad \Delta H_2 = +15{\cdot}15 \text{ kcal.} \quad (2)$$

(c) The heat of hydration of $\mathrm{Na_2CO_3}$

The right-hand members of equations (1) and (2) being identical, since solutions of the same composition are finally produced, we have, by subtracting (2) from (1),

$$\mathrm{Na_2CO_3 \ (solid)} + 10\mathrm{H_2O} = \mathrm{Na_2CO_3 . 10H_2O \ (solid)},$$
$$\Delta H_0 = -(6{\cdot}9 + 15{\cdot}15) = -22{\cdot}05 \text{ kcal.}$$

The results of the above examples may be compared with the following data extracted from information in *I.C.T.* 5, 202:

$$\Delta H_1 = -\ 6{\cdot}60 \text{ kcal. (exp. } a),$$
$$\Delta H_2 = +15{\cdot}21 \text{ kcal. (exp. } b),$$
$$\Delta H_0 = -21{\cdot}8 \ \text{ kcal. (exp. } c).$$

The following data from the same source may be quoted to serve as comparison if solutions of other compositions are employed in the experiments:

$$\mathrm{Na_2CO_3 . 10H_2O \ (solid)} + 390\mathrm{H_2O \ (liq.)} = \mathrm{Na_2CO_{3(400)}},$$
$$\Delta H = +16{\cdot}19 \text{ kcal.}$$
$$\mathrm{Na_2CO_3 \ (solid)} + x\mathrm{H_2O \ (liq.)} = \mathrm{Na_2CO_{3(x)}}.$$

x (g.-mol.)	$-\Delta H$ (kcal.)
400	5·62
200	5·93
100	6·35
50	7·00
30 (sat.)	7·57

It will be seen that the heat of dilution is of positive sign and of considerable magnitude.

In connexion with the above experiments it should be noted that the molar heat of solution ΔH may be defined in several ways, each of which corresponds to a characteristic and different value, for a given solute.

(1) *The total heat of solution* = the heat change when 1 g.-mol. is dissolved in a very large bulk of water (ΔH_{total} = 5·62 kcal. for Na_2CO_3).

(2) *The (integral) heat of solution* = the heat change when 1 g.-mol. is dissolved in just sufficient water to produce a saturated solution ($\Delta H_{sat.}$ = 7·57 kcal. for Na_2CO_3):

$$\Delta H_{total} = \Delta H_{sat.} + \text{the total heat of dilution.}$$

(3) *The heat change per g.-mol. when a small quantity of the solute is dissolved in the nearly saturated solution* ($\Delta H_{sol.}$). It is this last heat which is connected with the change of solubility with temperature, and it may be calculated from the temperature coefficient of solubility (see p. 67 and exp. *a* (iii), p. 70). $\Delta H_{sol.}$ for a salt may also be determined by measuring the heat change on relief of supersaturation.

To estimate in this way $\Delta H_{sol.}$ for $Na_2CO_3 . 10H_2O$ prepare in a thin-walled test-tube a specimen of the supersaturated solution described on p. 43. Close the tube with cotton-wool, and immerse it in the water of the calorimeter, the specimen of solution being below the level of the water outside. When steady temperature conditions have been reached add a fragment of the hydrate to the solution, and proceed to estimate the temperature rise in the usual way. For calculation the specific heat of the saturated solution may be taken as 0·80.

The heat of decomposition of hydrogen peroxide in aqueous solution

Prepare an approximately 1-volume ($N/5$) solution of the peroxide, and titrate a sample with standard potassium permanganate ($N/10$) to determine the exact concentration.

Place 75–100 ml. of this solution in the calorimeter, and establish the initial temperature line. Add about 2 g. of powdered manganese dioxide to act as catalyst, and stir well during the decomposition. Plot a cooling line as in previous experiments, and so obtain the corrected temperature rise (about 2°) and the heat of the reaction (q). The heat content of the catalyst and of the evolved oxygen may be neglected.

Calculate the heat of the reaction

$$H_2O_2 \,(aq.) = H_2O \,(liq.) + \tfrac{1}{2}O_2 \,(gas).$$

Assuming the equation

$$H_2 \,(gas) + \tfrac{1}{2}O_2 \,(gas) = H_2O \,(liq.), \quad \varDelta H = -68\cdot3 \text{ kcal.},$$

and neglecting the heat of solution of hydrogen peroxide, calculate its heat of formation from the gaseous elements.

(Giguère, Morissette, Olmos and Knop, *Can. J. Chem.* **33**, 804, (1955) find $\varDelta H_{25°} = -22\cdot62$ kcal, for the heat of decomposition of aqueous H_2O_2.)

Chapter VI

IONIZATION

General introduction

According to the theory put forward by Arrhenius (1887), the properties of a solution of a salt, such as potassium chloride, were to be referred to a partial electrolytic dissociation, resulting in a chemical equilibrium

$$KCl \rightleftharpoons K^+ + Cl^-.$$

The same extent or degree of dissociation (α) should be revealed by (1) measurements of electrolytic conductance, at a given dilution V, and at 'infinite' dilution:

$$\alpha_V = \frac{\Lambda_V}{\Lambda_\infty};\tag{1}$$

(2) the 'abnormality' in the freezing-point depression of aqueous solutions:

$$1 + \alpha = \frac{\Delta T_{\text{salt}}}{\Delta T_{\text{calc.}}} = i \text{ (van 't Hoff)}\tag{2}$$

($\Delta T_{\text{calc.}}$ is the depression calculated for a non-electrolyte at the concentration of the salt);

(3) measurement of the e.m.f. of concentration cells, which gives α from the relation

$$\ln \frac{\alpha_1 C_1}{C_2} = \frac{EF}{RT},\tag{3}$$

if C_2 is so low that $\alpha_2 = 1$ (for the derivation of this formula, see below, p. 193).

As experimental technique in applying these methods improved, it became increasingly evident that the values of α obtained from them for a given salt diverged in general far beyond the limits of experimental error, as the sample of data given in table 19 indicates.

Table 19. *Apparent degree of dissociation* (25°)

	C_{salt}	0·01	0·05	0·1	0·5	1·0	2·0	3·0	Coefficient
Sodium chloride	Λ_V/Λ_∞	0·936	0·882	0·852	0·773	0·741	—	—	β
	$(i-1)$	0·944	0·899	0·874	0·828	0·849	0·955	1·096	(α)
	e.m.f.	0·903	0·821	0·778	0·678	0·658	0·670	0·714	f
Hydrogen chloride	Λ_V/Λ_∞	0·972	0·944	0·925	0·890	0·845	—	—	β
	$(i-1)$	0·952	0·916	0·900	0·929	1·049	1·359	1·707	(α)
	e.m.f.	0·904	0·830	0·796	0·757	0·809	1·009	1·316	f

Such serious discrepancies, combined with the development of other independent methods of investigating electrolytes (such as the X-ray analysis of solid salts) and the increasing certainty of the conception that many simple ions have an atomic structure simulating that of an inert gas, have together caused the abandonment of the theory of partial chemical dissociation of salts in general.

The experimental methods 1, 2 and 3 above are now regarded as giving semi-independent fractional coefficients, reflecting the *physical* interaction of the ions of the salt, which is considered to be completely ionized at all concentrations. Thus the conductivity ratio $(\Lambda_V/\Lambda_\infty = \beta)$ gives principally a measure of the decrease of ionic mobility with increasing concentration. The osmotic coefficient, g (already evaluated for potassium nitrate in exp. *c*, p. 103), given by

$$g = \frac{i}{\nu} = \frac{1}{\nu}\frac{\Delta T_{salt}}{\Delta T_{calc.}} \quad \text{(see (2) above)}$$

(ν = the number of g.-ions from 1 g.-mol. of salt), shows the extent to which the ions interact in their effect upon the properties of the *solvent* (lowering its vapour pressure, i.e. decreasing its *activity*). Thus while the actual concentration of *each* ion of a salt like potassium chloride is always equal to that of the salt as a whole, $= C$, the *effective concentration* for calculating *osmotic effects* is only gC. The coefficient f derived

178 IONIZATION [CH.

from e.m.f. methods is related to the 'free energy' of the dissolved salt. If the ions of a salt composed of equally charged ions (potassium chloride or magnesium sulphate) were completely independent, the work expended, equal to the change of free energy, on transferring 1 g.-mol. of salt ($= 1$ g.-ion of each kind) from the concentration C_1 to the higher concentration C_2 would be $2RT \ln \dfrac{C_2}{C_1} = G_2 - G_1$. The interaction of the ions is revealed by the fact that the actual free energy change (measured by the e.m.f. of suitable cells—see below, p. 197) is found to be less, and given by $2RT \ln \dfrac{f_2 C_2}{f_1 C_1} (f_2 < f_1)$. The factor f is termed the *activity coefficient*, and $fC = a$, *the activity*, or *effective concentration* for calculating the *free energy* of the dissolved salt (for evaluation of f according to modern theory, see Chapter VII, p. 253).

All the three coefficients β, g and f tend towards unity as C becomes small, and it will be clear from the foregoing that the quantities $1 - \beta$, $1 - g$, and $1 - f$ each give a measure of the extent to which the ions fail to act independently, in connexion with definite kinds of phenomena in the solution. It has been shown theoretically (Debye and Hückel, and others) and confirmed by experiment that for simple salts of the type of potassium chloride a very simple relation exists (for dilute solutions) between these measures of 'imperfection' and the concentration, viz.

$$1 - y = A \sqrt{C}, \tag{4}$$

where $y = \beta$, g or f, and the constant A is different for each (more exactly $-\ln f = A \sqrt{C}$, but $-\ln f = -\ln[1 - (1-f)] = 1 - f$, when $f \simeq 1$: cf. Chapter VII, p. 253).

(A text-book of physical chemistry should be consulted for a detailed account of modern theories of strong electrolytes. Gurney, *Ions in Solution* (Camb. Univ. Press, 1936), provides a very clear exposition of Debye and Hückel's theory, and a stimulating discussion of strong electrolytes from the modern point of view.)

While the fundamental modification in the classical ionic

theory indicated above has proved necessary for *salts*, no such change has been needed in the general theory of the ionization of the acids and bases from which the salts are chemically derived. It is therefore possible to distinguish broadly two classes of electrolytes: (1) *strong electrolytes*, typified by salts, and a small group of acids and bases (the 'mineral' acids, and alkalis), to which the new theory applies; (2) *weak electrolytes*, typified by the majority of acids and bases, to which the practically unchanged theory of Arrhenius applies. It should be noted that this classification refers only to behaviour in *aqueous* solution and is not necessarily valid for other solvents; and that it is not rigid, for electrolytes of intermediate character exist. The distinction must ultimately be referable to special properties of the hydrogen and hydroxyl ions.

PART 1. THE CONDUCTANCE OF ELECTROLYTIC SOLUTIONS

The specific conductance κ of any conductor (metallic, electrolytic, etc.) is by definition numerically equal to the current (in amperes) flowing between two opposite faces of a cube of the conductor of 1 cm. side, when an e.m.f. of 1 V. is applied between those faces. The specific resistance ρ is the reciprocal of the specific conductance.

Let there be in 1 c.c. of a solution of a strong electrolyte formed of equally charged ions (e.g. potassium chloride, magnesium sulphate) c g.-*equivalents* of the electrolyte, and therefore c g.-*equivalents* of each ion. Then the number of g.-ions of each ion per c.c. is c/z, where z is the valency (or charge) of each ion. Also let $_{+}u_c$ and $_{-}u_c$ be the velocities of ionic drift (cm./sec.) in a field of 1 V./cm. length of the solution. The rate of electrical transport, in coulombs per second, i.e. *the current*, across a 1 cm. cube of the solution, when 1 V. is applied by means of suitable electrodes to opposite faces, is then

$$i = \frac{c}{z}zF\,(_{+}u + _{-}u)_c = \kappa, \text{ by the above definition,}$$

or $\kappa = c \, (_{+}l + _{-}l)_c$, where $l = uF$, the *mobility* of an ion. Hence $\kappa/c = \Sigma l_c$. The quantity $\kappa/c = \Lambda_c$ is termed the *equivalent conductance*. The mobility l is now often called the *equivalent conductance* for the ion concerned. $1/c = V$, the volume of the solution *in c.c.* containing 1 g.-equivalent of the electrolyte, is known as the *(equivalent) dilution*, and $\Lambda_c = \kappa V$.

At high dilutions the ions become physically independent of each other, and the mobilities reach limiting and characteristic maximum values l_∞.

$$\Lambda_\infty = \Sigma l_\infty \; [\text{Kohlrausch's law}].$$

The conductivity ratio β is $\Lambda_c/\Lambda_\infty = \dfrac{\Sigma l_c}{\Sigma l_\infty}$.

For an ion of given charge the mobility l_∞ is mainly determined by the hydration, which determines the size of the migrating ion, and by the viscosity of the solvent. As both hydration and viscosity decrease with rise of temperature, l_∞ and hence Λ_∞ increase rather rapidly under these conditions.

In solutions of weak, i.e. feebly ionized, electrolytes the ions are already at 'infinite' dilution and Σl may be taken as constant. The increase of Λ_c on dilution is to be attributed, not to an increase of mobility, but to an actual increase of ioniza-

Table 20. *Mobilities or ionic conductances at infinite dilution*

Ion	l_{25°	λ	Ion	l_{25°	λ
H^+	349·7	0·014	OH^-	197·8	0·016
Li^+	38·7	0·023	Cl^-	76·3	0·019
Na^+	50·1	0·021	Br^-	78·2	0·019
K^+	73·5	0·019	I^-	76·8	0·019
NH_4^+	73·5	—	NO_3^-	71·3	0·018
Ag^+	62·1	0·020	ClO_4^-	67·3	—
$\frac{1}{2}Mg^{2+}$	53·1	0·022	$\frac{1}{2}SO_4^{2-}$	80·0	0·020
$\frac{1}{2}Ca^{2+}$	59·5	0·021	$\frac{1}{2}C_2O_4^{2-}$	74·1	—
$\frac{1}{2}Ba^{2+}$	63·7	0·020	$C_2H_3O_2^-$	40·9	—

$$l_{t^\circ} = l_{25^\circ} \, [1 + \lambda(t - 25)]$$

tion. The concentration of each *ion* will be $\alpha_c c$, where α is the degree of ionization. Hence in this case

$$\kappa/c = \Lambda_c = \alpha_c [\Sigma l_\infty]$$

and $$\Lambda_c/\Lambda_\infty = \alpha_c.$$

The measurement of electrolytic conductance

The experimental method is almost always an application of the principle of Wheatstone's bridge, which gives directly the resistance. When four conductors are arranged to provide alternative paths to a current i (fig. 51), the relation

$$R_1/R_2 = R_3/R_4$$

must hold when no current passes through the detecting instrument I. In practice R_1/R_2 is made unity or a known ratio, R_3 is a resistance box, and R_4 the resistance to be deter-

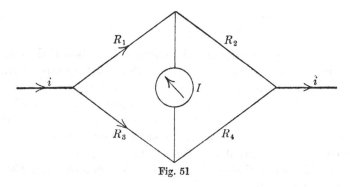

Fig. 51

mined. If R_4 is an electrolytic solution its resistance measured by using direct current will include not only its true ohmic resistance but also, as the result of electrolysis, resistance due to gas films at the electrodes, and an apparent resistance due to 'back e.m.f.' of the gas cell produced. These secondary effects can be minimized (1) by using alternating current, with telephone as detector I, (2) by platinizing the electrodes, i.e. coating them with an adherent layer of finely divided platinum. It is important that the alternating current should have a symmetrical (sine-wave) form, such as is produced by a thermionic oscillator.

The construction of a thermionic oscillator (fig. 52)

(1) *The equal inductances L_1 and L_2.* Wind $1\frac{1}{2}$ lb. of copper wire D.C.C. 22 S.W.G. on a 7 in. former for each coil. Fit the one coil above the other so that the turns of each are in the

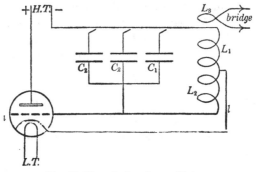

Fig. 52. Thermionic valve oscillator.

same sense, solder together the contiguous ends and the connexion l, and well cover the join with insulating tape. Then bind the two coils firmly together with more tape.

(2) *The condensers C_1 and C_2.* These must be of good ('wireless') quality; two (C_2) should have capacities of $0\cdot2\,\mu\text{F}$. and the third $0\cdot1\,\mu\text{F}$. The connexions should be so arranged that capacity from $0\cdot1$ to $0\cdot5\,\mu\text{F}$. can be put into the circuit.

(3) The valve V should be 6C4, which requires H.T. 100–250 V. and L.T. 6·3 V.

The Wheatstone bridge is fed by a small 7 in. coil L_3 (wind $\frac{1}{2}$ lb. of the above copper wire) acting as secondary of an air transformer to the main inductance $L_1 + L_2$.

Conductance cells

A great many forms of cell have been devised and used in particular investigations, especially in those of high precision, but the patterns shown in section in fig. 53 *a* and *b* serve well for

general purposes, including conductimetric titration. The two
circular platinum electrodes e (area about 2 cm.²) are firmly
attached by fusion of the glass (fig. 53c) to narrow-bore tubes,
which are filled with mercury to provide connexion from the
electrodes to the apparatus. The electrodes are rigidly fixed in
their relative positions by waxing the tubes into carefully
drilled holes in the ebonite cap c. For use in titration the containing vessel should have the (dipping) form of fig. 53b, and
it is convenient if the two vessels g_1 and g_2 are constructed from

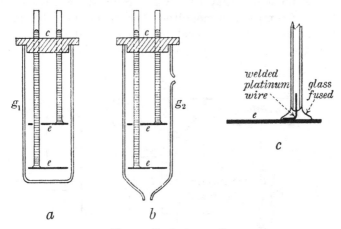

Fig. 53. Conductance cells.

tubes of the same diameter so that the cap c fits well in both,
and the same pair of electrodes can then be used in different
types of work.

The electrodes when affixed to the cap c are first cleaned by
immersion in hot concentrated nitric acid (or by standing
overnight in acid dichromate), washed well with distilled
water, and then supported in an inclined position (to allow
free escape of gas) in a solution prepared by dissolving 1 g. of
platinic chloride and 0·02 g. of lead acetate in 100 ml. of
dilute hydrochloric acid. The electrodes are connected to two
accumulator cells in series (4 V.), and electrolysis allowed to
proceed for 10–15 min. with frequent reversal of polarity. After

again washing the platinized electrodes in distilled water, the electrolysis with alternating polarity is continued in a solution of sodium sulphate, in order to displace material that may have been occluded in the surface during the platinizing. Finally, the prepared electrodes are washed for at least 2 hr. in frequently changed distilled water.

The Wheatstone bridge (fig. 54)

It can be shown that the bridge is most sensitive when current of comparable magnitude passes through each arm. For this reason the classical 'slide-wire' bridge may with advantage be replaced by the arrangement shown in fig. 54.

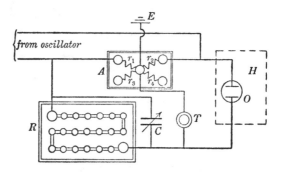

Fig. 54. Measurement of electrolytic conductance.

The ratio-box A can be constructed by mounting two 1000-ohm and two 100-ohm radio 'metallized' or wound-wire resistances on an ebonite panel. As the apparatus is always calibrated by taking the cell constant (see below), it is not essential that these resistances should be standard, but they should be constant. If a Post Office resistance box is available it will contain its own ratio arms, and the box A will not be required. The resistance box R may have a maximum of 10,000 ohms. The telephone T should have preferably two earpieces, and about 50 ohms resistance.

The sharpness of the sound minimum at the balance point is

usually improved by earthing the mid-point of the ratio arms
(*E*). Absolute silence at the minimum will only be obtained
when the capacity of the conductance cell is balanced as well
as its resistance. For this purpose a variable air-condenser *C*
(0·005 μF.) may be placed in parallel with *R*, but this refine-
ment is unnecessary for most ordinary work. For resistances
in the cell *O* up to 500 ohms use the 100-ohm ratio arms, and
the 1000-ohm arms for higher resistances. The specific resistance of electrolytic solutions decreases by 15–20 ohms per
degree, so that the accuracy attainable in measurement of
their conductance is likely to depend chiefly upon the nicety of
the regulation of the thermostat *H*.

The preparation of conductance water

The chief impurities in ordinary distilled water, viz.
ammonia and carbon dioxide, are removed by treatment with
alkaline permanganate, followed by distillation. A suitable

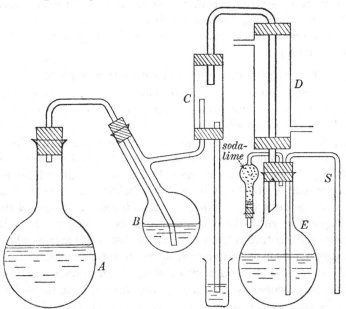

Fig. 55. Preparation of conductance water.

still, similar to an apparatus for steam distillation, is shown in
fig. 55. The inner tube of the condenser D must be of Jena,
Pyrex or silica glass. The turned-down joint C prevents contamination of the distillate by drainage from the bung of the
condenser tube. If a silica tube is used in the condenser a
straight type of joint C will be necessary, and the flask B
must then be of hard glass. To increase the speed of distillation
connecting tubes should be lagged.

Add to 2 l. of distilled water, contained in the flask A,
100 ml. of a saturated solution of potassium permanganate
and one stick of potassium hydroxide. Boil the solution
vigorously for 5–10 min. before connecting to the flask B.
After connecting, continue boiling, but allow the distillate to
condense in B until sufficient is present (50–100 ml.) to serve
as a trap for alkaline splash. Subsequently condensation in B
should be prevented by placing a small flame beneath. The
water should be received from the condenser in a hard glass
flask protected from recontamination, and should be siphoned
from the receiving flask as required.

(For review of methods, see Vogel and Jeffery, *J. Chem. Soc.*
1931, 1201.)

(a) *The equivalent conductance of magnesium sulphate.**

(i) *Determination of the cell constant.* Prepare from conductance water and pure (*Analar*) potassium chloride, previously powdered and dried in a steam oven, an accurately
$N/50$ or $N/100$ solution (assume for the formula weight of
potassium chloride the value adopted by Kohlrausch, viz.
74·59). Place enough of the solution in the cell to immerse
the upper electrode at least 1 cm. below the surface. This
should be done with a pipette so that the same volume of the
magnesium sulphate can afterwards be placed in the cell.
Take the electrodes from the bath of distilled water in which
they have been stored, remove as much water as possible from
the tubes and electrodes with a piece of filter paper, touching
for this purpose only the edges of the electrodes; immerse first

* Harkins and Paine, *J. Amer. Chem. Soc.* **41**, 1161 (1919).

in a spare quantity of the potassium chloride solution and then
set in position in the cell. The latter is now supported by a
holder in the thermostat and sufficient time allowed for temperature equilibrium to be established (10–15 min.). Connect
to the bridge and determine the resistance R as accurately as
the apparatus allows. (In determinations of high accuracy the
frequency of the current should be varied by changing the
circuit capacity in the oscillator. If the resistance appears to
shift slightly, it should be plotted against the frequency and
extrapolated to high frequency.)

The cell constant b is given by $b = R\kappa$. The specific conductances of potassium chloride solutions, as determined at
various temperatures by Kohlrausch, are given in table 21.
(Values of κ for solutions prepared with more detailed precautions are given by Grinnell Jones and Bradshaw, *J. Amer.
C.S.* **55**, 1780 (1933).)

Table 21. *Specific conductance* (κ) *of potassium chloride
solutions* (ohm^{-1})

Temp. °C.	N	$N/10$	$N/50$	$N/100$
10	0·0832	0·00934	0·001996	0·001019
18	0·0983	0·01120	0·002399	0·001224
25	0·1118	0·01289	0·002768	0·001412

(ii) Prepare an accurately $N/5$ solution of pure epsomite by
dissolving 24·65 g./l. of conductance water. Withdraw the
electrodes from the cell and wash them in distilled water.
Reject the solution of potassium chloride from the cell, which
is then thoroughly washed and dried, and then charged with
a volume of the $N/5$ magnesium sulphate solution equal to
that of the solution used in determining the cell constant.
Dry the electrodes and supporting tubes as described above,
immerse in a spare portion of the epsomite solution, and
finally set in position in the cell as before. Determine the
resistance (ρ about 100 ohms) when the solution is at the

thermostat temperature. Prepare, by diluting the solutions of epsomite with equal volumes of conductance water, further solutions of normality c respectively 0·1, 0·05, 0·025, 0·0125 and 0·00625. (If a *withdrawing* pipette is available, the reduction of concentration may be conveniently effected by removing half the contents of the cell and replacing with water, a store of which may be kept in the thermostat for this purpose.) Determine the resistance of each solution (ρ for the most dilute will be about 2000 ohms).

Tabulate the values of $b/cR = \kappa/c$ and plot against $\sqrt{c}$. Show that the plot becomes approximately linear for the solutions of the lower concentrations, in agreement with the formula

$$\Lambda_\infty - \Lambda_c = B\sqrt{c} \quad \text{(see equation (4), p. 178).}$$

Extrapolate the linear part to the Λ axis, and so estimate Λ_∞. (Ferguson and Vogel* have shown that a more exact relation for $MgSO_4$ is $\Lambda_\infty - \Lambda_c = Bc^{0·55}$ ($B = 626·2$).)

Compare the value estimated for Λ_∞ with that calculated from table 20, p. 180.

(b) *The dissociation (α) and the dissociation constant K of monochloracetic acid.*

(i) If it has not already been determined, determine the cell constant as under (a).

(ii) Prepare an approximately $N/10$ solution of the pure acid, in conductance water. Titrate a portion with standard baryta solution to determine the exact concentration. Determine the conductance at 25° of this solution ($\rho = $ about 200 ohms), and of a series of more dilute solutions (e.g. $C = 0·05$, 0·03, 0·02, 0·01 and 0·005), using the procedure described in detail in (a) above.

Tabulate $\Lambda_c = \kappa/C$ and $\Lambda_c/\Lambda_\infty = \alpha$, assuming $\Lambda_\infty = 389·6_{(25°)}$. Evaluate (1) $K_a = \dfrac{\alpha^2}{1-\alpha}C$ (use table 12, p. 108), and (2) K from $\log K = \log K_a - 1·014\sqrt{\alpha C}$. In this equation $K = f^2 . K_a$, and

* *Trans. Faraday Soc.* **23**, 406 (1927).

f is the mean activity coefficient of the ions (see p. 253). While K_a increases steadily with increasing concentration, K is constant ($1\cdot396 \times 10^{-3}$) for the range up to $0\cdot03\,\mathrm{N}$*†.

Graphical evaluation of Λ_∞ and K for weak acids (fig. 56)

From the dilution law in the form

$$\left(\frac{\Lambda_c}{\Lambda_\infty}\right)^2 = \frac{K}{C}\left[1 - \frac{\Lambda_c}{\Lambda_\infty}\right]$$

we find

$$\Lambda_c C = K\left[\frac{\Lambda_\infty^2}{\Lambda_c} - \Lambda_\infty\right].$$

Hence when $\Lambda_c C = \kappa$ is plotted against $1/\Lambda_c$ (fig. 56) a straight line results if K is constant. The intercept $OY = K\Lambda_\infty$, and $OX = 1/\Lambda_\infty$. Hence both K and Λ_∞ can be determined. This method is applicable to acids with K_a not greater than 10^{-5}.

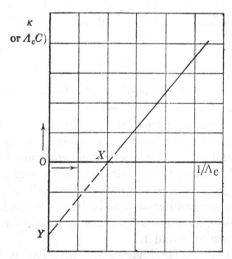

Fig. 56. Graphical evaluation of electrolytic dissociation constant.

* Saxton and Langer, *J. Amer. C.S.* **55**, 3638 (1933).
† Shedlovsky *et al.*, *Trans. Electrochem. Soc.* **66**, 176 (1934).

(c) *Conductometric titration.* *

When a strong acid is titrated with sodium or potassium hydroxide, the fast-moving H ion ($\varLambda \simeq 300$) is steadily replaced by the slower ion of the alkali metal ($\varLambda \simeq 50$), and the conductance of the solution therefore decreases rapidly as the alkali is added, until at the end-point S the solution contains Na^+ or K^+ and the anion of the acid in equivalent amounts. If the addition of alkali is continued beyond the end-point, the conductance rises again steeply as the fast OH ion accumulates ($\varLambda \simeq 180$). A plot of conductance against

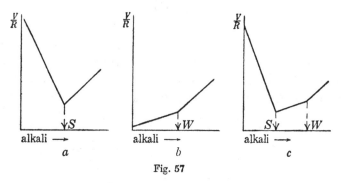

Fig. 57

alkali added takes the form of fig. 57 a, the end-point S being given by the sharp break in direction.

On the other hand, when a weak acid is titrated in the same way the conductance takes the course shown in fig. 57 b. The rise to the end-point W is due to the substitution of ionized salt for feebly ionized acid, and the steeper rise beyond the end-point again to the accumulation of OH ion. When a mixture of strong and weak acid is titrated a combination of a and b results, as in fig. 57 c, where the end-points for each acid are shown at S and W.

The method of conductometric titration is thus well adapted to the estimation of mixtures of acids of differing strengths, and to titration of single acids when for any reason

* Kolthoff and Sandell, *Quantitative Inorganic Analysis*: Macmillan, New York, 1952, p. 489.

indicators cannot be used (such as deep colour in the solution). The method can also be employed in many other volumetric estimations, e.g. the estimation of halides by titration with silver nitrate, and of sulphates by titration with baryta.*

Estimation of the amounts of hydrochloric and acetic acids present in an aqueous solution of the two acids.

To test the accuracy of the method prepare a solution of known composition by mixing $N/5$ hydrochloric acid and $N/5$ acetic acid in equal volumes. The dipping pattern of conductance cell is used (fig. 53, *b*, p. 183). The cell constant is

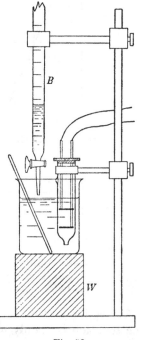

not required, nor is any strict thermostat control necessary. As there must not be large increases of volume during the titration, the alkali should be of not less than $N/2$ concentration, and be placed in a narrow-bore burette B (fig. 58). In the beaker put 20 ml. of the acid mixture; dilute and well mix with 100 ml. of distilled water (conductance water is here not necessary). Raise the beaker on the wood block W to immerse the electrodes (see that no air-bubbles lodge beneath either). Connect to the bridge, and determine the resistance to the nearest 1–2 ohms (ρ about 200 ohms). Run in from the burette 0·5 ml. of the alkali, receive any drop remaining on the burette tip on the glass rod, lower the beaker under the cell and well

Fig. 58

stir. Immerse the electrodes and allow the solution to drain from the cell once or twice by lowering the beaker, before again

* Harned, *J. Amer. Chem. Soc.* **39**, 252 (1917); Kolthoff, *Z. anal. Chem.* **61**, 229 (1922); *Amer. Abstr.* **16**, 2459 (1922).

observing the resistance, with the beaker in its former position. Proceed with the titration in this way until 10 ml. of the alkali have been added. Near the end-points NaOH = 4 ml. and 8 ml. add the alkali in smaller quantities of about 0·2 ml.

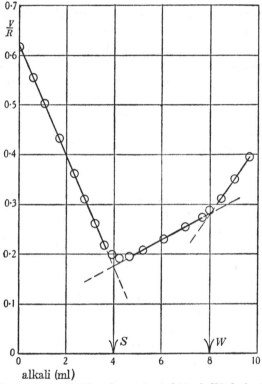

Fig. 59. *Conductimetric titration* of a mixture of 10 ml. $N/5$ hydrochloric acid and 10 ml. $N/5$ acetic acid with $N/2$ sodium hydroxide.

Correction for the relatively small change of volume during the titration is readily made by plotting not $1/R$, but V/R, against the volume of alkali added (V = total volume of solution in the beaker). In the immediate neighbourhood of the end-points the line is curved, and the true point is to be obtained by projection as in fig. 59.*

* Harms and Jahr, *Z. Elektrochem.* **41**, 130 (1935).

IONISATION (*continued*)

PART 2. ELECTROMOTIVE FORCE

In the experimental arrangement shown in fig. 60, two strips
(electrodes) of silver dip into solutions 1 and 2 each containing silver nitrate, and therefore Ag^+, at low concentrations

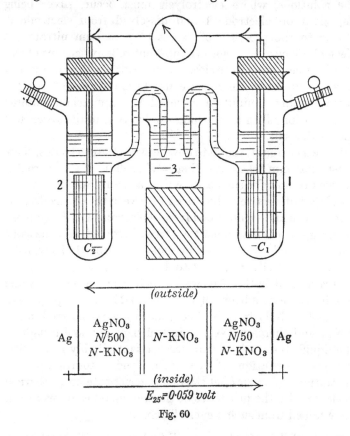

Fig. 60

C_1 and C_2 respectively (e.g. $C_1 = N/50$, $C_2 = N/500$). In
addition to the silver salt each solution contains potassium
nitrate in excess (e.g. normal concentration), and a solution 3

of potassium nitrate of the same concentration connects the
two electrode vessels. On joining the silver in 1 to that in 2
by a wire leading through a galvanometer, it will be found that
a current (of 'positive' electricity) flows from 1 to 2, i.e.
from the side of higher concentration C_1 to that of the lower C_2.
It is evident that this current is part of a circuit, and current of
equal strength must be flowing in the reverse direction through
the solutions, where electrolysis must occur, silver being
deposited on electrode 1 and dissolved from electrode 2.
Owing to the presence of excess of potassium nitrate the
electrolytic current is borne almost entirely by the ions of this
salt between the electrodes. Hence neither the increase of
silver produced by electrolytic dissolution in vessel 2, nor the
corresponding diminution in vessel 1, is appreciably altered
by migration of Ag ions, which would occur in the absence of
the carrier salt.

Since a current is found to flow between the electrodes, there
must be an e.m.f. between them $= E$ (volts). If the current is
allowed to flow for such a time that a small fraction x of one
faraday F of electricity has passed between the electrodes, the
work that could be obtained from the current is EFx (joules).
In 2, x g.-ion of silver must then have dissolved, and the same
amount have been removed by deposition from 1. To transfer
the dissolved silver from 2 to 1 and thus restore the concentrations of both solutions to their original values would
require the expenditure of work $W = xRT \ln (C_1/C_2)$ (joules).
x g.-mol. of NO_3^- have in theory also to be transferred, but if
KNO_3 and therefore NO_3^- is in sufficient excess throughout
the liquid, the work of transference is negligibly small. The
successive operations of current flow and restoration of concentrations, *if carried out reversibly*, constitute an isothermal
cycle, and by the principles of thermodynamics no work can
be obtained from such a cycle. Therefore

$$EFx = xRT \ln (C_1/C_2), \quad \text{or} \quad E \text{ (volts)} = (RT/F) \ln (C_1/C_2).$$

It will be clear that this formula will apply generally to all such
arrangements of metal electrodes placed in dilute solutions of

their salts, in which they can dissolve electrolytically. For a metal of valency z and ion M^{z+} the general formula becomes

$$E = \frac{RT}{zF} \ln \frac{C_1}{C_2}. \tag{1}$$

Introducing the values of R (in joules), F in coulombs, taking $T = 298° = 25°$ C. and changing to common logarithms, we find

$$E_{25°} = \frac{0 \cdot 059}{z} \log \frac{C_1}{C_2}. \tag{2}$$

The e.m.f. given by this formula is necessarily the reversible or maximum e.m.f., as determined by a compensation method (see below, p. 198). Formula (2) has been fully confirmed by experiment (see exp. a, p. 221).

The original arrangement is called a *concentration cell*, and it will be noted that any such cell with a z-valent metal working at 25° develops a (maximum) e.m.f. of $59/z$ mV., when the ratio $C_1/C_2 = 10$; the *absolute* magnitudes of the concentrations are irrelevant. (In general the work W expended in transferring 1 g.-mol. of silver nitrate from the concentration C_2 to the higher concentration C_1 would be $W = RT \ln(a_1/a_2)$, where a is the *effective* concentration or *activity* of the salt (p. 178, above). If f is the activity coefficient, $a = fC$, and $\frac{a_1}{a_2} = \frac{f_1 C_1}{f_2 C_2}$; in the arrangement described above the presence of excess potassium nitrate at the same concentration on each side ensures that $f_1 = f_2$, although neither $= 1$; see Chapter VII, p. 253.)

While the above thermodynamic reasoning is very valuable in facilitating a calculation of the e.m.f. without any assumption about how it arises, the latter question remains of fundamental importance, and demands some discussion. A metal may be regarded as an assemblage of kations M^{z+} and electrons $z\mathbf{e}$, and in being composed of two oppositely charged units it has a formal analogy to a solid electrolyte such as a salt, but with the fundamental difference from our present point of view that only the kations are soluble in water. Their solubility, although inaccessible to experimental determination for the reasons to be given, may be taken to have the definite value C_0 at the temperature T. If then a metal is brought into contact with a solution of its salt in which the concentration of kations C differs from C_0, the metal ions tend to dissolve if $C < C_0$, and to be deposited if $C > C_0$. However, only *incipient* dissolution or deposition is possible, since the free dissolution of kations from the metal, or their deposition from solution, is prevented by the opposite charges (electrons or anions) remaining on the metal or in the solution. Dynamic equilibrium is reached by the metal assuming a charge of sign and magnitude depending on the sign and magnitude of the difference $C - C_0$, the concentration C undergoing only negligible change. We see in particular that the position of this metal-solution equilibrium, measured by the charge on the metal, is conditioned solely by the relation of C to C_0, that is by C, as C_0 must be assumed constant for a metal at constant temperature. Ions of salts other than those of the metal concerned only have influence in so far as they may alter the effective value of the concentration C.

It will now be clear why current flows when the electrodes of a concentration cell are connected by a wire, for when insulated they must have different charges, since $C_1 \neq C_2$. Metallic connexion allows the otherwise impossible dissolution and deposition of kations, by providing a path through the external circuit for the charges on the electrodes. Current flows until $C_1 = C_2$, when the electrodes become equally charged. This line of reasoning leads at once to the conception of single electrode potentials, for if we imagine unit positive charge travelling from the solution into the metal (electrode 1) it will yield or perform work W according as the metal is negative or positive, and $W = E_1$, the electric potential between electrode 1 and the solution. E_1 is itself not strictly an e.m.f., as it is associated with an electrically static condition. The potential E_1 also depends only on C_1. If E_2 is the corresponding potential at the second electrode, then the e.m.f. of the whole cell is

$$E = E_1 - E_2 = \frac{RT}{F} \ln a_1 - \frac{RT}{F} \ln a_2 \quad \text{(p. 195)}.$$

Hence $E_{1,2} = \dfrac{RT}{F} \ln a_{1,2} + E^0$, where E^0 is a constant for a given metal or other kathode-forming element such as hydrogen.

Unfortunately there is no generally accepted method for determining single electrode potentials, and thus to find E^0. It is therefore customary to set E^0 arbitrarily $= 0$ for hydrogen electrodes, that is, the potential of the hydrogen electrode in H^+ ion of effective concentration $a_H = 1$ (standard hydrogen electrode) is taken as zero, since $\log 1 = 0$. On combining this standard hydrogen electrode (H) with other electrodes, or 'half-cells' (M), containing metals in contact with solutions of their salts (under definite conditions, e.g. saturation, normal solution, or solution of unit activity), the total measured e.m.f. of the resulting *voltaic* cell,

$$\mathrm{H_2} \left| a_H^{\pm} = 1 \right\| \begin{array}{c} \text{Metallic salt} \\ \text{solution} \end{array} \left| M, \right.$$

is
$$E = E_H^0 - E_M = 0 - E_M.$$

Any electrode potential E_M can be thus evaluated in an arbitrary but self-consistent way. Some electrode potentials of practical importance are listed in table 22, p. 219. The e.m.f. to be expected from a voltaic combination is expressible in a general formula of the type

$$E_{25^\circ} = \frac{0 \cdot 059}{z} \log \frac{a_1}{a_2} + (E_1^0 - E_2^0).$$

For a simple concentration cell $E_1^0 = E_2^0$, and the formula reduces to (2) above, under conditions in which $\dfrac{a_1}{a_2} = \dfrac{C_1}{C_2}$.

From the thermodynamic standpoint the quantity EzF gives the maximum work liberated at constant temperature and pressure by the chemical changes taking place in the cell and giving rise to the e.m.f. $= E$. Thus EzF gives the change of free energy (ΔG) associated with the cell reaction, and measurements of e.m.f. form one of the most accurate and accessible means of evaluating such free energy changes. As there is usually no difficulty in determining the temperature coefficient of the e.m.f. $= dE/dT$, the heat change at constant pressure (change of heat content) can also be found by employing the general thermodynamic relation

$$\Delta G - \Delta H = T \left(\frac{\partial \Delta G}{\partial T} \right)$$

(see Chapter v, Thermochemistry, pp. 156, 157), which in this case takes the form

$$EzF - \Delta H = TzF \left(\frac{dE}{dT} \right)_p .$$

The principle of the measurement of e.m.f. by the potentiometer, as first proposed by Latimer Clark, is represented in fig. 61. OR is a stretched wire of uniform thickness and con-

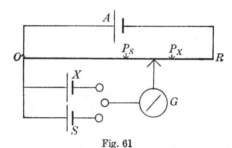

Fig. 61

siderable resistance, similar to that used in the measurement of resistance by the method of Wheatstone's bridge. A battery A maintains between O and R a steady e.m.f., which need not be known, but must be greater than E_X or E_S to be evaluated. Similar poles of the cells S and X and the battery are connected at O. With S in circuit, the sliding contact C is moved to a point P_S such that no deflexion is seen in the galvanometer G; with X in circuit the corresponding position is P_X. Then

$$\frac{E_S}{E_X} = \frac{OP_S}{OP_X} .$$

If S is a cell of standard e.m.f., E_X can be immediately calculated.

In practice a simple straight wire cannot be given sufficient resistance, and in accurate work must be replaced by a much longer wire wound on a revolving drum, or by two resistance boxes, as shown in plan in fig. 62.* The resistance boxes 1 and 2 should preferably be similar, each having the denominations 1, 2, 2, 5, 10, 20, 20, 50, 100, 200, 200 and 500 ohms, making a total of 1110 ohms in each box. *A* is a lead accumulator cell, in a partially discharged state, so that its e.m.f. is steady at

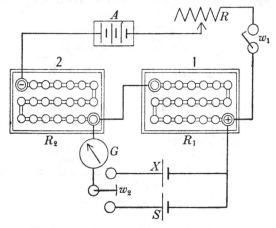

Fig. 62. Measurement of electromotive force.

approximately 2·0 V., *S* is a standard cell (for its preparation see below), and *X* the cell whose e.m.f. is to be measured. For all ordinary work a sensitive needle galvanometer (*G*) is satisfactory; a capillary electrometer is not to be recommended, and a mirror instrument is excessively sensitive. It is convenient, but not essential, to have an adjustable (sliding) resistance *R*, of maximum value 1500–2000 ohms. The switches w_1 and w_2 should be of the mercury cup type, and are easily constructed as follows. Short sections of quill glass tubing are drawn out and the narrowed portion bent round as

* Cf. Lewis, Brighton and Sebastian, *J. Amer. Chem. Soc.* **39**, 2246 (1917).

shown in fig. 63. The cups are fixed by heating their bases into a block of paraffin wax. A small amount of mercury is placed in each cup, and connecting wires l (with ends amalgamated) are inserted in the narrow projections. Contact between the mercury in the cups is made by means of a bent section of stout copper wire (also amalgamated) supported by a short piece of ebonite or wood (e) which serves as a handle in operating the switch. A two-way switch (w_2) is best made by fixing three cups at the corners of a *scalene* triangle, and providing

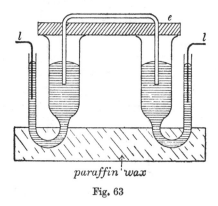

Fig. 63

two connecting pieces of different lengths, to avoid the possibility of inadvertently closing the switch in the wrong direction.

In the following explanation of the use of the potentiometer, we shall suppose that a cadmium cell ($E = 1\cdot018$ V.) is available as the standard S. Out of box 1 take plugs to give a total of 1018 ohms in the box, and out of box 2 plugs to give 92 ohms resistance, so that the boxes together have a total resistance ($R_1 + R_2$) of 1110 ohms, the value one alone would have if all the plugs were removed. Set the resistance R at about its mid-point, close w_1, and *momentarily* switch in the standard cell S at w_2. Observe the direction and magnitude of the galvanometer deflexion. Increase the resistance R and again momentarily switch in S, and note whether the galvanometer swing is lessened, increased or reversed. Continue to adjust R and make tests by closing w_2, until the galvanometer ceases to be deflected in either direction. On no account must the standard cell be switched in for more than a very short time, until it is nearly balanced.

Open the switch w_1, replace all plugs in box 1, and remove all those of box 2. Take plugs out of 1 to give a resistance R_1 = about 1000 E_X, where E_X is the suspected voltage of X; replace in 2 the plugs of the denominations removed from 1, so that the total resistance $R_1 + R_2$ is again 1110 ohms. Close w_1, bring X into circuit by closing w_2 in its direction for a few seconds, and observe the movement of the galvanometer needle. Adjust the resistances R_1 and R_2 in the two boxes, *plugs removed from the one being always replaced in the other*, thus keeping the total resistance constant at 1110 ohms, until a balance is obtained, indicated by the absence of deflexion on closing w_2. The resistance R remains at the value to which it was set with the cell S. E_X in volts is then given by $R_1/1000$. This mode of operation allows the required voltage to be directly read in millivolts numerically equal to R_1.

If a resistance R (adjustable or fixed) is not available, the standardization should proceed along the same lines as the second part of the observations above. Thus if a balance is obtained when S is in circuit, with $R_1 = 565$ ($R_2 = 545$), and when X is in circuit, with $R_1 = 340$ ($R_2 = 770$),

$$E_X = \frac{340}{565} \times 1\cdot018 \text{ V.}$$

Greater accuracy can be obtained by this method of working if a fixed resistance of about 1000 ohms can be placed at R.

Standard cells

A cell to be used as a standard of e.m.f. should be (*a*) readily reproducible from well-defined materials, (*b*) chemically and physically stable over long periods when out of use, (*c*) as far as possible unpolarizable. In addition to these essential qualities it is very desirable that its e.m.f. should have a low-temperature coefficient. The cadmium-amalgam cell (Weston cadmium cell)

$$\text{Cd-Hg} \left| \begin{array}{c} \text{CdSO}_4 \text{ satd.} \\ \text{HgSO}_4 \text{ satd.} \end{array} \right| \text{Hg}$$

is generally held to satisfy these various requirements better than any other yet devised.

Preparation of a standard cadmium cell

The usual form of glass containing vessel, with fused-in platinum wires, is shown, with suitable dimensions, in fig. 64a.

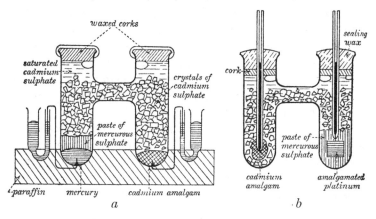

Fig. 64. Cadmium amalgam standard cells.

An alternative form, avoiding wires at the base, is described later. The vessel should be thoroughly cleaned and dried before being filled as directed below. (In order to secure perfect contact between the platinum wires and the mercury or amalgam electrodes, it is advantageous to amalgamate the internal projections of the wires. For this purpose a small amount of a solution of mercurous nitrate is poured into each of the limbs in turn and electrolysis conducted with the fused-in wire as kathode and a spare platinum wire, dipped into the solution, as anode. The cell is thoroughly washed with dilute nitric acid and then distilled water to remove all mercury salt, and dried.)

(a) *Pure mercury.*

Place in a stout glass-stoppered bottle about 20 ml. of commercially pure mercury, and 30 ml. of a solution of mercurous nitrate made by dissolving 1 g. of the salt in 100 ml. of water to which has been added six drops of concentrated nitric acid. Shake vigorously (on a machine) for half an hour. Decant the solution, add another 30 ml. of the solution of mercurous nitrate, and repeat the shaking. After a second decantation thoroughly wash the mercury with successive portions of distilled water, pour into a porcelain basin, and remove the greater part of the water by absorption in pieces of filter paper. Finally, remove the last traces of water by filtering through a paper folded as usual and pierced with a small hole at the angle.

(b) *Saturated cadmium sulphate solution.*

(Solubility at $23° = 72.5$ g. $CdSO_4 . \frac{8}{3}H_2O$ in 100 g. water.)

First powder a weighed amount of the pure sulphate (e.g. *Analar* brand) in a mortar; add an equal weight of distilled water in portions while continuing the grinding. Hot water should not be used as basic salts are likely to be precipitated, and in any case the solubility changes very little with temperature. After half an hour allow the undissolved sulphate to settle and decant the solution.

(c) *Cadmium amalgam.*

Dissolve a weighed piece of pure metal (about 2 g.) in seven times its weight of pure mercury, previously heated to about 100°. Pour the amalgam before it solidifies on cooling into one limb of the H-shaped vessel, so as to cover the fused-in wire with a thick layer. (The amalgam produced as above contains 12·5 % of cadmium, but no special care need be taken to ensure this exact composition, as a variation from 8 to 13 % produces only a negligible change in the e.m.f. of the cell,

if used at temperatures near 20°. Amalgams with less than
10 % of cadmium remain liquid at room temperature.)

(d) *Mercurous sulphate paste.*

A quantity of granular mercurous sulphate should be freed
from soluble (mercuric) sulphate by shaking with three successive portions of the saturated cadmium sulphate solution
(not with water). Finally, decant as much of the solution as
possible (or filter on a Buchner funnel). The moist sulphate is
at once transferred to a mortar, a few drops of pure mercury
and also of the cadmium sulphate solution are added, and the
mixture is rubbed into a uniform grey paste. Put into the
remaining limb of the cell pure mercury to form a thick layer
over the fused-in wire, and then pour in sufficient of the
sulphate paste to form a layer not less than 5 mm. thick over
the mercury. After this has been done great care must be taken
that the cell is not shaken and the sulphate allowed to intrude
between the mercury electrode and the contact wire. Both
limbs of the cell are now loosely filled with crystals of cadmium
sulphate, and then the saturated solution of this salt is poured
in so as nearly to fill the vessel. The cell is closed with cork
disks, which are then coated with paraffin wax. An air-bubble
should remain under each disk to allow for expansion. The
terminals of the cell must be well insulated from each other.
This is ensured by fixing it into a block of paraffin wax and
passing the lead wires into mercury cup terminals, as shown in
fig. 64. The practice of attaching the cell to a wooden base-
board, with screwed-in terminals, is very strongly to be de-
precated. The prepared cell should be allowed to stand for
2–3 days to acquire a constant e.m.f.

A more robust and also very well-insulated form of construction is shown in fig. 64 b. The cadmium amalgam is cast on
to a capillary tube with fused-in platinum wire by immersing
the tip of the tube in the still liquid amalgam. The other
(mercury) electrode is formed by welding a small piece of
platinum foil to a fused-in platinum wire, and amalgamating
with mercury by electrolysis in mercurous nitrate solution as

described above. The paste of mercurous sulphate, which is poured round this electrode, hardens after a short time so that the cell can even be inverted without damage. The e.m.f. of the cadmium cell has been determined with great exactness and is given by the formula

$$E_{t^o} = 1 \cdot 01827 - 0 \cdot 0000406\,(t - 20)\ \text{V (international)}.$$

It has been suggested that the stability of the cell is improved if the solution of cadmium sulphate is made up in dilute (e.g. $0 \cdot 1\,N$) sulphuric acid. The increase of e.m.f. due to the presence of acid is very slight, and is given by $\Delta E = 0 \cdot 000855 \times$ (normality of the acid).

The Helmholtz cell

A fairly serviceable standard cell, of simpler construction than the cadmium cell, is the combination

$$\text{Zn amalgd.} \left| \begin{array}{c} \text{ZnCl}_2 \text{ solution} \\ \text{Hg}_2\text{Cl}_2 \text{ sat.} \end{array} \right| \text{Hg.}$$

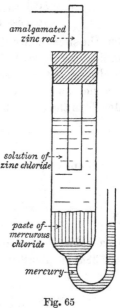

This cell may conveniently take the form shown in fig. 65. The solution of zinc chloride is prepared as follows: To 100 ml. of pure concentrated hydrochloric acid (*Analar*, $d = 1 \cdot 16$), contained in a flask, add in portions 50 g. in all of pure zinc oxide (*Analar*), while cooling the solution from time to time in tap water. When oxide remains undissolved cork the flask and well shake the liquid. Filter through a pleated paper. Determine the specific gravity ($1 \cdot 4$–$1 \cdot 5$) of the filtered and cooled solution either with a hydrometer, or by weighing a 10 ml. portion. A paste

Fig. 65

of calomel is then prepared as for a calomel electrode (see below), substituting the solution of zinc chloride made as above for the solution of potassium chloride. A rod of pure zinc is well amalgamated by rubbing with a solution of mercurous nitrate, thoroughly washed with distilled water, and then placed in position as anode.

The data of Labendzinski* show that the e.m.f. of this cell depends on the specific gravity of the zinc chloride solution (between 1·75 and 1·39), according to the linear formula

$$E_{15°} = 1·3423 - 0·245 \text{ (sp. gr.)}_{15°}.$$

The temperature coefficient is $+0·00009$.

(The useful 'life' of the cell is said to be limited by the tendency of the zinc pole to become covered with a film.)

Hydrogen electrodes

It is found by experiment that the arrangement

$$\text{Pt, H}_2 \left| \begin{array}{c} \text{HCl } (C_1) \\ \text{excess KCl} \end{array} \right| \begin{array}{c} \text{KCl} \\ \text{solution} \end{array} \left| \begin{array}{c} \text{HCl } (C_2) \\ \text{excess KCl} \end{array} \right| \text{Pt, H}_2$$

(cf. the silver concentration cell, p. 193 above), in which Pt, H_2 indicates a piece of platinized foil in contact with pure gaseous hydrogen (at atmospheric pressure), acts in a quite normal manner as a concentration cell, reversible to H ion, and develops the e.m.f. to be expected, viz.

$$E_{25°} = 0·059 \log \frac{a_2}{a_1} = 0·059 \log \frac{C_2}{C_1}.$$

When one of the electrodes is standard, i.e. $a_2 = 1$.

$$E_{25°} = -0·059 \log a_1$$

or
$$pH_{25°} = \frac{E}{0·059}. \tag{1}$$

If a calomel electrode (e.g. the decinormal electrode, $E_{25°} = 0·334$) is substituted for the standard H electrode,

$$pH_{25°} = \frac{E - 0·334}{0·059}. \tag{2}$$

Thus from equations (1) or (2) the pH of a solution can be directly determined with an accuracy equal to that with which the e.m.f. can be measured. All other methods of

* Z. Elektrochem. **10**, 78 (1904).

measuring pH must be regarded as depending ultimately on this electrical method.

Liquid junction potentials

In general at the junction between solutions containing different electrolytes a boundary potential is set up due to the unequal mobilities of the ions present; thus at the junctions

$$\text{Na}^+ \leftarrow \| -\text{H}^+ \quad \text{or} \quad \text{Na}^+ \quad \| \quad \text{Na}^+$$
$$\text{Cl}^- \quad \| \quad \text{Cl}^- \quad \quad \text{Cl}^- \leftarrow \| -\text{OH}^-$$
$$\dashv\!\!-\!\!- \quad \quad \quad -\!\!-\!\!\vdash$$
$$\pi \quad \quad \quad \quad \pi$$

the high mobility of the H or OH ion relative to the mobilities of the other ions results in a strong tendency for these ions to diffuse to the left, and a diffusion potential π is established in the direction shown. These potentials may amount to as much as 50 mV., but, owing to their sensitiveness to the conditions at the junction, are very difficult to calculate. They may be minimized, usually to values negligible in ordinary work, by connecting half-cells with a concentrated (e.g. saturated) solution of a salt whose ions have about equal mobilities, e.g. potassium chloride or ammonium nitrate.*

Conventions in regard to cell diagrams, sign of electrode potential, etc.

Most of the conventions summarized below have already been exemplified in the preceding pages, and will be consistently used in the experimental section.

A *single vertical line* | represents the junction of an electrode (usually metallic) and a solution, e.g. Zn | ZnSO$_4$ sat.

A *double vertical line* ‖ represents a junction between two solutions, e.g. 0·1 KCl ‖ KCl sat. ‖ 1·0 KCl.

A *long arrow*, embracing the whole cell diagram, gives the direction of the current *inside* the cell. A *short line* with indication of positive end (⊣—) gives the direction of a potential difference (electrode or liquid junction potential).

* For a detailed discussion of these potentials see Glasstone, *The Electrochemistry of Solutions*, Methuen, 1945, pp. 314–20.

The letter E, alone or with left-hand subscript, e.g. $_lE$, $_rE$, indicates e.m.f. (mnemonic: emF, leFt); E, with a right-hand subscript, e.g. E_2, indicates an *electrode potential* (mnemonic: electRode, Right). The sign ^{0}, e.g. $E^0_{Cl^-}$, will only be used when the *activity* of the reversible ion in solution is unity: the potential is then termed a *standard* potential.

Liquid junction potential will be represented, when required, by the letter π. All electrode potentials will be assumed to be measured on the hydrogen scale ($a_H = 1$).

Sign of electrode potential

An electrode potential is reckoned positive when the electrode is positive to the solution with which it is in equilibrium.

Examples of the use of these conventions

Cell diagrams:

(1)
$$E = 0\cdot376$$

$$\text{Hg} \left| \begin{array}{c} \text{Hg}_2\text{SO}_4 \\ N\ \text{H}_2\text{SO}_4 \end{array} \right\| N\ \text{KCl} \left\| \begin{array}{c} \text{Hg}_2\text{Cl}_2 \\ N\ \text{KCl} \end{array} \right| \text{Hg}$$

$$E_m = 0\cdot679 \quad \pi = 0\cdot023 \qquad E_c = 0\cdot280$$

(2)
$$E = 0\cdot3567$$

$$\text{Pt, H}_2 \left| \begin{array}{c} 0\cdot1\ N\ \text{HCl} \\ 0\cdot9\ N\ \text{KCl} \end{array} \right\| \begin{array}{c} \text{KCl} \\ \text{sat.} \end{array} \left\| \begin{array}{c} 0\cdot1\ N\ \text{KCl} \\ \text{AgCl} \end{array} \right| \text{Ag}$$

$$E_H = 0\cdot0692 \qquad \pi\ \text{zero}\ \ \pi\ \text{zero} \qquad E_{Ag} = 0\cdot2875$$

Electrode potentials (at 25°):

$$\text{Zn} \,|\, N\ \text{ZnCl}_2 \qquad\qquad \text{Zn} \,|\, \text{Zn}^{+2}$$
$$E_{Zn} = -0\cdot777 \qquad\qquad E^0_{Zn} = -0\cdot762$$

$$\text{Ag} \left| \begin{array}{c} \text{AgCl} \\ \text{Cl}^- \end{array} \right. \qquad\qquad \text{Ag} \left| \text{Ag}^+ \right.$$

$$E^0 = 0\cdot2223 \qquad\qquad E^0_{Ag} = 0\cdot799$$

(reversible ion chlorine)   (reversible ion silver)

*The construction of hydrogen electrodes**

(1) *The gas-bubbling type* (fig. 66).

A glass shield with side-tube *s* and expanded lower portion
B is first prepared. At *O* a small hole (diameter 2 mm.) is
made, by inserting a platinum wire into the locally melted
glass, and drawing out. When the elec-
trode is in use (exp. *a*, p. 221) the
hydrogen should be released through
this hole and not from the bottom
of the tube, so that the electrode is
alternately bathed in gas and the solution. It is well to test whether the tube
will operate in this way before proceeding further with the construction.
A piece of platinum foil, preferably
not less than 1 cm.² in area, is welded
to a platinum wire. (Lay the foil flat
on a charcoal block, hold the wire above
it with tongs; heat the wire and foil
simultaneously to bright redness with
the foot blowpipe, and while both are
hot *gently* tap the wire into contact
with the foil with a *light* hammer. The
wire should be long enough to cover
the breadth of the foil and project about
¼ in. on one side.) Attach a bead of

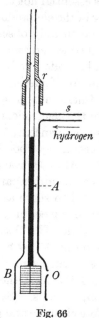

Fig. 66

cobalt glass ('schmeltz-glas') symmetrically to the wire by
fusion, and then fuse into the end of a tube *A*, which should
be 2–3 in. longer than the shield *B*, and of calibre that will
pass freely into its narrow part. (During the fusing it is
important that the wire should remain fairly symmetrically
embedded in the cobalt glass, and not be allowed to slip side-
ways to touch and fuse to the ordinary glass. The free end of the

* For bibliography see Glasstone, *op. cit.* on p. 207, pp. 376–7.

wire should project well into the tube. The joint should be well annealed in the luminous flame after completion.) The foil, after being cleaned by standing (overnight) in acid dichromate, or more quickly by immersing in hot concentrated nitric acid, is washed and *lightly* platinized by electrolysis, as recommended for electrodes of conductance cells (p. 183); contact is made by running a small amount of mercury into the tube A. It is essential not to platinize too heavily; in the final appearance of the electrode the original gleam of the platinum should *not* be entirely obscured. The treated electrode is washed in distilled water, and then made several times alternately anode and kathode in dilute sulphuric acid, with intervals of about a minute of gassing between the changes of polarity; the final treatment must be as kathode. If gas does not arise uniformly from the electrode, it is not clean, and the preparation should be recommenced. A spare piece of platinum foil must be used as second electrode in the platinizing and in the acid electrolysis. After thoroughly washing the electrode with distilled water, the upper part of the tube A is moistened with aqueous glycerol to serve as lubricant, and the tube passed upwards through a piece of stout rubber tubing r previously attached to the shield. The electrode is set in its final position under the shield (fig. 66) by holding and drawing upon the *upper* part of tube A when a sufficient length has been projected through r. Avoid touching the electrode with the fingers during these manipulations. (Setting the electrode in its final position before cleaning and platinizing is not to be recommended, as it is then usually difficult to secure an even coating of deposit.) The electrode should not be allowed to dry after its final preparation, and it should be kept immersed in distilled water, or dilute sulphuric acid, when not in use.

(2) *The enclosed type.*

The type of electrode just described is essential for electrometric titrations and some other purposes, but for the determination of the pH of a single solution of fixed composition it is unnecessarily wasteful of hydrogen. Enclosed types of electrodes are shown in fig. 67 A, B below. The platinized electrode (*e*) is prepared as before. The vessel is first completely filled with the solution under examination, and the bung carrying the electrode and tapped tubes 1 and 2 then set in place with tap 2 open, so that a little of the solution issues from it, and no air bubble is left under the bung. In type B the expanded part of the ground joint *j* should be raised so that

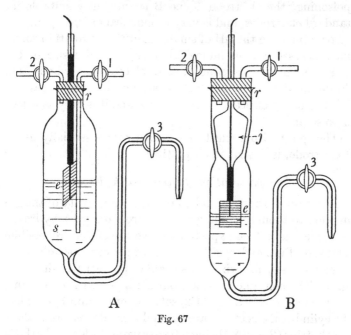

Fig. 67

the joint does *not* engage when the bung is inserted. The supply of hydrogen is now attached at 1, tap 2 shut, and 1 and 3 opened, so that hydrogen displaces solution to a level such that the electrode is about half immersed. In type A the tap 3 is then shut, and 2 opened, hydrogen thus continuing to bubble through the solution for some time to remove dissolved air. With type B after shutting 3 the joint *j* is closed by a slight rotatory depression of the electrode tube through the bung. The vessel is now supported on a wheel, which is slowly rotated (by a motor) for half an hour, when the oxygen due to dissolved air is completely removed by catalytic combination with hydrogen on the electrode. After this treatment the electrode gives a very stable and constant potential for long periods. This method of removing oxygen should not be attempted with the simpler type A, as the contact between the rubber bung and the solution will probably result in 'poisoning' the electrode. Type B is eminently suitable for standard electrodes, and is very economical of hydrogen.

For estimating the pH of small quantities of solution sometimes necessitated in biological work, etc., the size of the apparatus must be reduced, and electrodes formed from platinum wire are often employed, but in such cases the stability of the electrode is rather low, and it is very sensitive to poisoning.

(The quinhydrone electrode, which acts essentially as an H electrode, is described on p. 218.)

The preparation of electrolytic hydrogen

Unless precautions are taken to ensure the purity of the gas supplied to them (including freedom from oxygen), hydrogen electrodes cannot be expected to give the stable (reversible) potential. Pure hydrogen may be most conveniently obtained by electrolysis, for which a suitable apparatus is shown in fig. 68. The container G is a stout glass jar, with cork bung, and the electrolyte a 15–20 % solution of sodium hydroxide. The cylindrical electrodes are cut and formed from stout sheet nickel, strips (S_1 and S_2) being left to serve as leads. The kathode is contained inside the lower expanded part of an inner glass tube K, and its lead sealed through the top of the tube with wax: the anode is held in the position shown in the figure, by passing its lead tightly through the cork bung. The auxiliary kathode N is made by rolling fine nickel gauze, bending the roll into a ring and attaching a nickel wire or strip as lead, which is also to be carried through the bung. The gauze ring is set closely round the lower circumference of the bell-shaped kathode vessel, which should be supported in the cork about

5 mm. above the base of the container. The gauze kathode is connected to the main kathode through a resistance of 40–50 ohms.

When the generator is working the fine bubbles of hydrogen originating from the gauze electrode effectively 'scrub' the

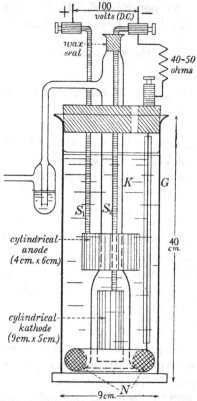

Fig. 68. Electrolytic preparation of hydrogen.

liquid near the bottom of the kathode bell free from dissolved oxygen, and prevent contamination of the hydrogen with this gas. A current of 4–5 amp. supplies gas at a rate and of purity suitable for direct use in a bubbling electrode. The low resistance of the generator prevents any trouble arising from overheating.

Reference electrodes

In determining the potential E_1 of an electrode (such as that of an H electrode in pH measurement, equation (2), p. 206), it is necessary to combine it with a second or reference electrode of known potential E_r. If the e.m.f. of the combination is E, E_1 is known from the relation $E = E_1 - E_r$.

Calomel electrodes

These are of the type

$$\text{Hg} \left| \begin{array}{l} \text{Hg}_2\text{Cl}_2 \text{ (solid + saturated solution)} \\ \text{KCl (conc.} = C) \end{array} \right.$$

Such electrodes act in the normal way through the reversible change $\text{Hg} \rightleftharpoons \text{Hg}^+ + \text{e}$ (more probably $2\text{Hg} \rightleftharpoons \text{Hg}_2^{++} + 2\text{e}$), but owing to the operation of the principle of the solubility product,

$$a_{\text{Hg}^+} = \frac{\text{const.}}{a_{\text{Cl}^-}},$$

and therefore the electrode may be regarded as (indirectly) reversible to chloride ion. The e.m.f. of a combination of two calomel electrodes with different concentrations of potassium chloride is given by the usual type of formula:

$$E = \frac{RT}{F} \ln \frac{(a_1)_{\text{Hg}^+}}{(a_2)_{\text{Hg}^+}} = -\frac{RT}{F} \ln \frac{(a_1)_{\text{Cl}^-}}{(a_2)_{\text{Cl}^-}}.$$

Reference electrodes are usually prepared in three forms, depending on the concentration of the potassium chloride: the *normal* electrode, $[\text{KCl}] = M$, the *decinormal* electrode, $[\text{KCl}] = 0.1\,M$, and the *saturated* electrode, with a saturated solution of potassium chloride. (For the potentials of these electrodes, see table 22, p. 219.)

Preparation of calomel electrodes.*

The electrode may be conveniently set up in a glass bottle, of about 100 ml. capacity (fig. 70). Pure mercury (see p. 203) is

* Cf. Lewis, Brighton and Sebastian, loc. cit. p. 199.

first placed in the thoroughly cleaned bottle to give a layer about 1 cm. deep. A tube *t* containing mercury and with fused-in platinum wire at its lower end to serve as contact with the mercury of the electrode is then introduced (the wire may with advantage be previously amalgamated as described on p. 202). The tube should contain sufficient mercury to sink the platinum contact to rest on the base of the bottle.

'Pure' calomel to which a few drops of mercury have been added is shaken in a spare bottle with two successive portions of the prepared solution of potassium chloride to be used in the electrode, to remove traces of mercuric compounds; the solution is decanted and rejected before the second portion is added. The calomel is treated a third time in a similar way, but the solution of potassium chloride is retained after decantation, for the final filling of the cell. The pasty calomel is carefully poured on to the mercury in the electrode bottle, great care being taken that no liquid paste intrudes between the contact wire and the mercury. (Grinding the calomel in a mortar with the potassium chloride solution and mercury, which is sometimes advocated, is not to be recommended.*) If a saturated electrode is being prepared, crystals of pure potassium chloride should be added to the paste after it has been poured into the bottle. The rubber stopper, carrying the connecting tube with tap 1 and another short tube terminating in a rubber tube and pinch-clip 2, and bored to receive the tube *t*, is now set firmly in place; the tube *t* must not be lifted out of the mercury. The electrode is completed by drawing into it, through the connecting tube by suction at 2, a quantity of the solution of potassium chloride reserved from the third washing above. The electrode should be allowed to stand for a day or two before use.

* See Lewis and Sargent, *J. Amer. Chem. Soc.* **31**, 362 (1909).

The silver-silver chloride electrode

Owing to its simplicity and robustness this type of electrode has superseded the calomel electrode in much recent work.* For ordinary work (accuracy 1 mV.) the following preparation suffices: A strip of pure (*not* 'standard') silver is cut to the form shown in fig. 69, and freed from grease by washing with

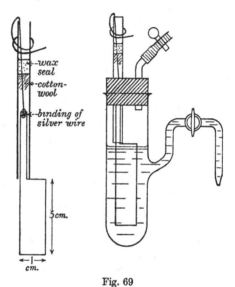

Fig. 69

acetone. After a short annealing by passing through a Bunsen flame a connecting silver wire is attached by binding into a small hole, bored at the extremity of the narrow part, which is then inserted into a length of glass tubing, and the connecting wire bent over the top of the tube so as to retain the silver strip firmly in place. The strip is now again washed with acetone, and coated with a film of chloride by electrolysis as anode in normal hydrochloric acid with a current of 4–6 mA.

* For a critical account of preparation and performance, see Smith and Taylor, *Natl. Bur. Stand. J. Res.* **20**, 837 (1938), and Brown, *J. Amer. Chem. Soc.* **56**, 646 (1934).

for about an hour; a piece of platinum foil is used as kathode. The electrode should finally have a *uniform* purplish colour.

The electrode may be set up for use in a vessel of the form shown in fig. 69. The electrolyte may be molar or decimolar potassium chloride or hydrochloric acid. (For potentials see table 22, p. 219.)

The quinhydrone electrode

The quinhydrone electrode (see p. 218) is perhaps the most readily set up of all reference electrodes. The form with $0.01 N$ HCl and $0.09 N$ KCl is recommended by Veibel* and by Hovorka and Dearing† as superior in stability and reproducibility to the calomel electrode. For potentials see table 22, p. 219.

Oxidation-reduction (redox) cells

The possible equilibrium $Fe^{2+} \rightleftharpoons Fe^{3+} + e$, where e stands for free electrons, cannot be established in a homogeneous solution of ferric and ferrous salts, owing to the impossibility of the existence of free electrons in solution. A piece of inert metal, such as platinum, placed in the solution will provide a supply of electrons, but they will be available only at the metal-solution interface, where a local equilibrium will be set up. An application of the law of chemical equilibrium shows that the concentration of electrons at the metal surface is given by

$$[e] = k \frac{[Fe^{2+}]}{[Fe^{3+}]} = kr, \tag{1}$$

and this concentration will in general be different from the normal concentration when the metal is out of contact with the solution; in other words, the metal assumes a charge of electricity and a definite potential difference will exist between it and the solution, determined in magnitude by the value of the right-hand member of (1).

It will be recalled that in the ordinary electrode equilibrium metal-metal salt (p. 196), it is the concentration C of the

* *J. Chem. Soc.* **123**, 2206 (1923).
† *J. Amer. Chem. Soc.* **56**, 243 (1934).

metallic ions which determines the value of the electrode charge and the potential is given by $E = \dfrac{RT}{F} \ln C + E^0$. In an analogous manner the redox potential is given by

$$E_r = \frac{RT}{F} \ln kr + E_r^0.$$

We may therefore set up a redox concentration cell

$$\mathrm{Pt} \left| \begin{array}{c} \dfrac{\mathrm{Fe}^{2+}}{\mathrm{Fe}^{3+}} = r_1 \\[2mm] [(\mathrm{NH_4})_2\mathrm{SO_4}] \end{array} \right\| (\mathrm{NH_4})_2\,\mathrm{SO_4} \left\| \begin{array}{c} \dfrac{\mathrm{Fe}^{2+}}{\mathrm{Fe}^{3+}} = r_2 \\[2mm] [(\mathrm{NH_4})_2\mathrm{SO_4}] \end{array} \right| \mathrm{Pt,}$$

of which the e.m.f. should be (cf. p. 193)

$$E = \frac{RT}{F} \ln \frac{r_2}{r_1}.$$

This result is completely confirmed by experiment. For redox systems in which n valency stages are concerned (e.g. for $\mathrm{Sn}^{2+} \rightleftharpoons \mathrm{Sn}^{4+} + 2\mathrm{e}, n = 2$)

$$E = \frac{RT}{nF} \ln \frac{r_2}{r_1}.$$

Redox cells with different redox systems on the two sides (e.g. Fe^{2+}, Fe^{3+} and Ce^{3+}, Ce^{4+}) correspond to voltaic cells, and if r is made equal on the two sides, the e.m.f. of such combinations is given by

$$E = \frac{RT}{F} \ln \frac{k_2}{k_1} + \text{constant.}$$

In the example above the direction of the e.m.f. shows that electrons flow from the Fe side to the Ce side, and the magnitude of the e.m.f. is a measure of the power of the Ce system to oxidize the Fe system. A study of the e.m.f. of such systems has given much valuable quantitative information on oxidizing and reducing powers of reagents.

The quinhydrone electrode

The organic redox system hydroquinone-quinone is concerned with the equilibrium

$$\text{Quinone} + 2\mathrm{H}^+ + 2\mathrm{e} \rightleftharpoons \text{Hydroquinone,}$$

from which as before we find

$$[e]^2 = K \frac{[H_2Q]}{[Q]} \cdot \frac{1}{[H^+]^2} \quad (H_2Q = \text{hydroquinone}, \ Q = \text{quinone}).$$

If we arrange that $[H_2Q] = [Q]$, $[e] = \dfrac{\text{constant}}{[H^+]}$. In this case

Table 22. *Important electrode potentials*

(1) Hydrogen and quinhydrone electrodes

Pt, H₂	(acid) $a_H = 1$ ($C_{HCl} = 1.18\,N$)	$E_H^0 = 0$, *at all temperatures*
Pt, H₂	N HCl ($a_H = 0.810$)	$E_{25°} = -0.0054$
Pt	acid, with quinhydrone ($a_H = 1$)	$E_t^0 = 0.7175 - 0.00074t$* (between 0 and 37°)
Pt	quinhydrone $0.01\,N$ HCl $0.09\,N$ KCl	$E_{18°} = 0.5826$†

(2) Calomel electrodes

C_{KCl}	$E_{25°}$	Temp. coefficient
0·1	0·334‡	0·00007§
1·0	0·281‡	0·00024§
sat. (4·1 N)	0·242‡	0·00066‖
C_{HCl}		
0·1	0·3345‡	
1·0	0·2676‡	

* Biilmann and Krarup, *J. Chem. Soc.* **125**, 1954 (1924).

† Veibel, *J. Chem. Soc.* **123**, 2203 (1923); Hovorka and Dearing, *J. Amer. Chem. Soc.* **56**, 243 (1934); *ibid.* **57**, 446 (1935).

‡ Randall and Young, *J. Amer. Chem. Soc.* **50**, 989 (1928).

§ Glasstone, *The Electrochemistry of Solutions*, 1945, p. 327.

‖ Vellinger, *Arch. phys. biol.* **5**, 119 (1927); *Chem. Soc. Abstr. A* (1928), p. 369.

Table 22 (*continued*)

(3) Other mercury electrodes

Hg $\left\vert\begin{array}{l}\text{Hg}_2\text{SO}_4 \text{ sat.}\\ N \text{ H}_2\text{SO}_4\end{array}\right.$	$E_t = 0.6778 + 0.00028\,(t-18)*$
Hg $\left\vert\begin{array}{l}\text{HgO}\\ N \text{ KOH}\end{array}\right.$	$E_t = 0.1100 - 0.00011\,(t-25)\dagger$
Hg $\left\vert\begin{array}{l}\text{HgO}\\ N \text{ NaOH}\end{array}\right.$	$E_t = 0.1135 - 0.00011\,(t-25)\dagger$
Hg $\left\vert\begin{array}{l}\text{HgO}\\ N/10 \text{ NaOH}\end{array}\right.$	$E_t = 0.1690 + 0.00007\,(t-25)\dagger$ (E_t for above electrodes with alkali based on $E_{18°}$ for normal calomel electrode $= 0.283$)

(4) Silver-silver chloride electrodes

Ag $\left\vert\begin{array}{l}\text{AgCl}\\ \text{Cl} -\end{array}\right.$ $(a_{\text{Cl}}=1)$	$E_{25°} = 0.2223 - 0.00065\,(t-25)\ddagger$
Ag $\left\vert\begin{array}{l}\text{AgCl}\\ \text{KCl } 0.1\,N\end{array}\right.$	$E_{25°} = 0.2875\ddagger$

* Glover, *J. Chem. Soc.* (1933), p. 10.
† Donnan and Allmand, *J. Chem. Soc.* **99**, 845 (1911).
‡ Bates, *Electrometric pH Determination*, Wiley, New York, 1954, p. 208.

then the redox potential is determined solely by the concentration of hydrogen ion; the redox electrode is therefore equivalent to a hydrogen electrode.

It happens that quinone and hydroquinone form a compound, quinhydrone, in which the two constituents combine in equimolecular proportions. The condition $[H_2Q] = [Q]$ is thus automatically satisfied when quinhydrone is dissolved in the solution.

Typical experiments involving measurements of e.m.f.

N.B. In all experiments on e.m.f. not only the value of the e.m.f. but also the direction of the current (*inside* the cell) must be carefully noted. The direction is to be ascertained from the polarity of the potentiometer connexion required to give a null point. It is desirable that the potentiometer should be permanently labelled with the polarity given by its working battery (*A*, fig. 61, p. 198). The immediate results of an e.m.f. determination are most compactly and informatively recorded by drawing a cell diagram (see notes, p. 207).

(*a*) *The e.m.f. of concentration cells.*

(i) Set up in turn the combinations

$$\text{Pt, H}_2 \left|\begin{array}{c} 0{\cdot}1\,N\text{ HCl} \\ 0{\cdot}9\,N\text{ KCl} \end{array}\right\| \begin{array}{c} \text{KCl} \\ \text{sat.} \end{array} \left\|\begin{array}{c} N\text{ KCl} \\ \text{Hg}_2\text{Cl}_2 \end{array}\right| \text{Hg} \qquad (1)$$

and
$$\text{Pt, H}_2 \left|\begin{array}{c} 0{\cdot}01\,N\text{ HCl} \\ 0{\cdot}9\,N\text{ KCl} \end{array}\right\| \begin{array}{c} \text{KCl} \\ \text{sat.} \end{array} \left\|\begin{array}{c} N\text{ KCl} \\ \text{Hg}_2\text{Cl}_2 \end{array}\right| \text{Hg} \qquad (2)$$

as shown in fig. 70. To obtain the acid solutions for the H electrodes make up 20 ml. of N or $N/10$ HCl in a 200 ml.

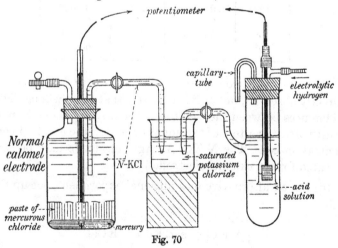

Fig. 70

graduated flask with normal KCl solution. Hydrogen should be allowed to pass through the electrode shields for at least 15 min. before a reading of e.m.f. is attempted. Immediately before a reading is to be made shut off the gas supply, leaving the electrode half immersed in the solution. The bubbling should then be resumed, and further readings taken to confirm the constancy of the e.m.f. Unless a very sensitive form of galvanometer is available, the taps of the electrode vessels should be opened during a reading: before opening see that the levels of solution in the two vessels and in the connecting beaker are equal, to avoid siphoning. Assume the temperature of the measurements to be room temperature.

Deduce from the observations the e.m.f. and the direction of the current in the combination

$$\text{Pt, H}_2 \left| \begin{array}{c} 0\cdot1 \text{ HCl} \\ 0\cdot9 \text{ KCl} \end{array} \right\| \begin{array}{c} \text{KCl} \\ \text{sat.} \end{array} \left\| \begin{array}{c} 0\cdot01 \text{ HCl} \\ 0\cdot9 \text{ KCl} \end{array} \right| \text{Pt, H}_2. \qquad (3)$$

Check the value with the following table (see equation (1), p. 195):

2·303 RT/F		2·303 RT/F	
16°	0·0573	21°	0·0583
17°	0·0575	22°	0·0585
18°	0·0577	23°	0·0587
19°	0·0579	24°	0·0589
20°	0·0581	25°	0·0591

(If two H electrodes are available the e.m.f. (3) may be directly observed, but as it is probable that there will be slight differences between the electrodes, the method above is more trustworthy as a test of the equation giving the e.m.f. of concentration cells. The N KCl Ag-AgCl electrode may be substituted for the calomel electrode.)

(ii) As an alternative or additional experiment, set up the cell

$$\text{Ag} \left| \begin{array}{c} \text{AgCl} \\ N \text{ KCl} \end{array} \right\| \begin{array}{c} \text{KCl} \\ \text{sat.} \end{array} \left\| \begin{array}{c} \text{AgCl} \\ 0\cdot1 \, N \text{ KCl} \end{array} \right| \text{Ag,}$$

using Ag-AgCl electrodes prepared as described on p. 216. Compare the direction of the current in relation to the concentration of Cl^- with that in exp. (i).

(b) *The solubility of silver chloride.*

Set up and determine the e.m.f. of the cell

$$\text{Ag} \left| \begin{array}{c} 0 \cdot 1 \, N \, \text{AgNO}_3 \\ N \, \text{NH}_4\text{NO}_3 \end{array} \right\| \begin{array}{c} \text{NH}_4\text{NO}_3 \\ \text{sat.} \end{array} \right\| \begin{array}{c} \text{AgCl} \\ N \, \text{KCl} \end{array} \right| \text{Ag.}$$

Prepare the silver electrodes as described on p. 216, but omit the 'chloridizing' treatment for that on the silver nitrate side. Assuming the e.m.f. is given by the formula

$$E = \frac{RT}{F} \ln \left[\frac{C_2}{C_1} \right]_{\text{Ag}^+} = 2 \cdot 303 \, \frac{RT}{F} \log \left[\frac{C_2}{C_1} \right],$$

calculate the solubility C_1 of silver chloride in normal potassium chloride. (Use the table of values of RT/F on p. 222.)

Using the relation $C_1 = C_0^2 / a_{\text{Cl}^-}$, calculate the solubility C_0 in water. (Take $f\,[\text{Cl}^-] = a_{\text{Cl}^-}$ as 0·60 in N KCl.)

(C_0 should be approx. 0·0014 g./l. at room temperature.)

(c) *The determination of the pH of solutions.*

To determine the pH of the standard acetate buffer, and the dissociation constant of acetic acid (K). Titrate 20 ml. of an approximately $N/5$ solution of acetic acid with $N/5$ sodium hydroxide, using phenolphthalein. To prepare the standard buffer (acid = salt) add to a volume of the acid *half* the volume of the sodium hydroxide required for complete neutralization.

Determine the e.m.f. (E) of a combination with the normal or decinormal calomel or silver-silver chloride electrodes similar to that of exp. (i), using for the buffer solution preferably one of the enclosed types of H electrode described on p. 211. The connecting electrolyte should be saturated potassium chloride.

Calculate the pH of the buffer from the formula

$$p\text{H} = -\log a_{\text{H}^+} = \frac{E - E_r}{0 \cdot 1983 T} \quad (E \text{ and } E_r \text{ in } \textit{millivolts}).$$

E_r is the electrode potential of the reference electrode, which should be ascertained from table 22 (pp. 219–20). The factor $0\cdot1983 = \dfrac{2\cdot303R}{F} \times 1000$. For most ordinary work this factor may be put $= 0\cdot200$, and the formula takes the easily remembered form $pH = \dfrac{E - E_r}{0\cdot2T}$.

From the equation for a buffer solution (p. 230)

$$pH = pK - \log \frac{\text{acid}}{\text{salt}},$$

it will be seen that for the standard buffer

$$pH = pK = -\log K.$$

Redox, or oxidation-reduction, potentials

(a) Prepare solutions approx. $N/10$ in respect to oxidizing or reducing properties as follows:

Potassium dichromate: $4\cdot9$ g. of the salt in 1 l. of (a) approx. normal potassium chloride, (b) normal hydrochloric acid.

Stannous chloride: $11\cdot3$ g. of the hydrated salt ($SnCl_2 . 2H_2O$) per litre, made up (a) in normal potassium hydroxide, (b) in normal hydrochloric acid.

Sodium sulphite: $12\cdot6$ of the hydrated crystals per litre of normal potassium chloride.

Ferrous ammonium sulphate: $39\cdot2$ g. of the salt per litre, made up (a) in water with only sufficient sulphuric acid to prevent hydrolysis, (b) in normal sulphuric acid.

Determine approximately the e.m.f. of the redox cell

$$\text{Pt} \left| \begin{matrix} K_2Cr_2O_7 \\ \text{or } H_2Cr_2O_7 \end{matrix} \right\| \begin{matrix} KCl \\ \text{sat.} \end{matrix} \left\| \begin{matrix} \text{reducing} \\ \text{solution} \end{matrix} \right| \text{Pt}$$

formed by taking the reducing solutions prepared as above in turn. Clean platinum wires, fused into supporting glass tubes to provide mercury connexion to the apparatus, will suffice as electrodes, which should be completely immersed in the solutions contained in electrode vessels similar to that for the bubbling H electrode in fig. 70.

At room temperature the results should correspond with the following redox potentials in volts:

$H_2Cr_2O_7$	1·12	$FeSO_4$	0·35
($K_2Cr_2O_7$ + acid)		Na_2SO_3	0·30
$K_2Cr_2O_7$	0·78	$SnCl_2$ + HCl	0·22
$FeSO_4 + H_2SO_4$	0·51	Alkali stannite	− 0·58
H_2SO_3	0·44	($SnCl_2$ + alk.)	
(Na_2SO_3 + acid)			

Thus for $H_2Cr_2O_7/FeSO_4$ the e.m.f. should be

$$1·12 - 0·35 = 0·77 \text{ V.},$$

and for $H_2Cr_2O_7$/stannite $1·12 + 0·58 = 1·70$ V.

Although these e.m.f., first measured by Bancroft*, are not strictly stable and reversible, they provide a useful comparison of redox power.

The pronounced influence of H-ion concentration upon the redox potential or strength will be noted.

(b) *The quinhydrone electrode*†. To prepare the solution, well shake 100 ml. of the acid liquid with about 1 g. of solid quinhydrone in a *glass*-stoppered bottle. Then pour the solution so formed together with some of the undissolved solid into an electrode vessel as used for H electrodes (fig. 67 A, B, p. 211). A small piece of platinum foil, mounted on a glass tube as for hydrogen electrodes (p. 209) *but not platinized*, is used as metallic electrode. It should be kept in acid dichromate when out of use, and well washed before being completely immersed in the solution treated with quinhydrone. *The quinhydrone electrode may only be used in liquids of pH less than 6.*

(i) The experiments a (i) and c, p. 221, may be repeated, substituting the quinhydrone electrode for the gas electrode.

(ii) Determine the standard potential E^0 of the quinhydrone electrode by measuring the e.m.f. of the cell

$$\text{Pt} \left| \begin{array}{c} \text{quinhydrone} \\ \text{0·1 } N \text{ HCl} \\ \text{0·9 } N \text{ KCl} \end{array} \right\| \begin{array}{c} \text{KCl} \\ \text{sat.} \end{array} \left\| \begin{array}{c} \text{0·1 } N \text{ HCl} \\ \text{0·9 } N \text{ KCl} \end{array} \right| \text{Pt, H}_2.$$

* *Z. physikal Chem.* **10**, 387 (1892).

† Biilmann and Krarup, *J. Chem. Soc.* **125**, 1954 (1924). See also references, p. 219.

It will be seen that e.m.f. of this cell, $= E_q^0$, is independent of the composition of the electrolyte.

Compare the value of E^0 obtained with that given in the table of electrode potentials on p. 219.

Potentiometric titration

(a) *Acid-alkali.*

Arrange apparatus as in fig. 71, with bubbling H electrode,

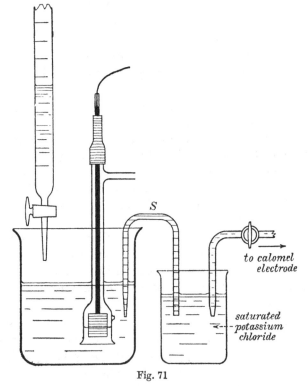

Fig. 71

and a calomel or Ag-AgCl reference electrode. As it is inadvisable to allow the connecting tube of a standard electrode to dip directly into the solution being titrated, electrolytic connexion is conveniently made by the use of an agar-KCl bridge (S, fig. 71), prepared as follows: Heat on a water-bath until dissolved 3 g. of agar with a solution of 40 g. of potassium chloride in 100 ml. of water. Fill a bent tube S, of which one end has been drawn to a fine point, with the still warm solution and leave with the point dipping into the solution until the solution has set to a gel. If at the wider end of the tube the gel has shrunk away from the end, cut the tube so that it is quite filled with the gel. The bridge when out of use should be kept immersed in saturated chloride.

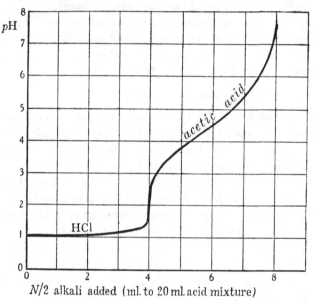

Fig. 72. *Potentiometric titration* of a mixture of 10 ml. $N/5$ hydrochloric acid and 10 ml. $N/5$ acetic acid with $N/2$ sodium hydroxide (cf. fig. 59).

(i) As an example of the method of titration the mixture of hydrochloric and acetic acids used in exp. *c*, p. 190, may be titrated with $N/10$ sodium hydroxide (in this method changes of volume during titration are relatively immaterial). The e.m.f. is measured on the potentiometer after each addition of alkali. It will usually not be quite steady, owing chiefly to the inevitable presence of dissolved oxygen in the liquid, but only quite approximate values of the e.m.f. are required (to

2–3 mV.). The pointed end of the agar bridge is to be immersed in the solution undergoing titration, but only when a reading is being made.

Plot the calculated pH (see formula on p. 224) against the volume of alkali added, when a curve of the form shown in fig. 72 should be obtained.

(*Note.* Should the electrode be in a deteriorated condition it will probably be found to yield more regular results in acid-alkali titration if the acid is placed in the burette.)

(ii) Titrate a standard solution of potassium dihydrogen phosphate with sodium hydroxide, and so confirm the evidence of the polybasicity of phosphoric acid obtained by a thermo-chemical method in exp. *b*, p. 164.

(b) *Oxidation-reduction titrations.**

The necessary apparatus is similar to that of (a), but the H electrode is replaced by a plain platinum wire dipping into the solution undergoing titration: the wire should be cleaned just before use by heating to redness in an *alcohol* flame.

As an example of the use of the method a solution of ferrous ammonium sulphate in dilute sulphuric acid may be titrated with $(N/10)$ potassium dichromate. The sudden and large jump of potential (about 300 mV.) at the end-point is due to the sudden change of the solution from reducing to oxidizing character at that point. The method is adapted to the titration of very dilute solutions, such as are found for example in estimating iron in glass.

* Lingane, *Electroanalytical Chemistry*, Interscience Publishers, New York, 1953.

IONIZATION (*continued*)

PART 3. HYDROGEN-ION INDICATORS AND BUFFER SOLUTIONS

(1) Indicator constants

The chemical action of an indicator is summarized in the equation $Y + H^+ \rightleftharpoons R$, whence, by applying the law of equilibrium, we obtain

$$[H^+] = k_i \frac{R}{Y}, \quad \text{or} \quad pH = pK - \log \frac{R}{Y}. \tag{1}$$

In these equations k_i is the equilibrium (or indicator) constant, and R and Y represent respectively the 'acid' and 'alkaline' forms of the indicator; $pK = -\log k_i$. When an indicator is used that exhibits distinct colours in both alkaline and acid solutions (e.g. methyl orange, bromthymol blue, etc.), the *tint*, determined by the ratio R/Y, is dependent only upon the pH of the solution, and not on the amount of indicator present, since a further addition will cause an increase in R and Y in the same ratio. With one-colour indicators (e.g. phenolphthalein, p-nitrophenol, etc.) the eye discerns only the presence of the one coloured form (R or Y), and a further addition of indicator will be seen to increase the *depth* of the colour with no change in tint, although of course the amount of the colourless form is increased in the correct proportion. There is thus a fundamental difference in the use of the two classes of indicators in quantitative work. With a two-colour indicator any convenient amount of indicator may be used, as, in dilute solutions, the *tint* is unalterable provided the pH remains constant: with a one-colour indicator the quantity of indicator must be carefully equalized in solutions under comparison, and only then will the *depth* of colour give a measure of pH.

Although the ratio R/Y must alter continuously throughout the whole possible range of pH values, the eye fails to notice further colour change outside the approximate range R/Y = 10 to 0·1. The working range of an indicator is thus between pH values given by $pH = pK \pm 1$, and the indicator is most sensitive near $pH = pK$. The experimental determination of the indicator constant k_i (or pK) is based on the fact, deducible from equation (1), that when $R = Y$, $pH = pK$.

(2) Buffer solutions

In a solution containing hydrogen ions and the anions of a weak acid there will exist the chemical equilibrium

$$H^+ + A^- \rightleftharpoons HA,$$

whence, by the law of equilibrium,

$$[H^+] = K_a \frac{\text{acid}}{\text{anion}}. \tag{2}$$

If the solution contains not only the free acid but an appreciable amount of a (completely ionized) salt of the acid, the anions, which actually originate from both acid and salt, may be regarded as due only to the salt, and equation (2) becomes

$$\left.\begin{aligned} [H^+] &= K_a \frac{\text{acid}}{\text{salt}}, \\ pH &= pK - \log \frac{\text{acid}}{\text{salt}}. \end{aligned}\right\} \tag{3}$$

The preparation of buffer solutions is founded on this equation, as a definite ratio of acid to salt establishes a definite (and approximately calculable) pH, which is not seriously changed by dilution (see p. 236) or by addition of small amounts of free acid or base. The resistance of the solution to such addition is a maximum for acid = salt, when $pH = pK$, and falls off on each side, so that the working range of a buffer mixture is usually taken as given by $pH = pK \pm 1$ (cf. the range of an indicator). To cover the complete range of pH values encountered in practice it is therefore necessary to employ a series of acids, of increasing K_a.

Equations (1) for indicators, and (3) for buffers, being of the same form, may be exhibited graphically in the same diagram (fig. 73), where ordinates give either the relation of acid to salt, or of R to Y. Further, the same characteristic S-shaped line is the (titration) curve for the neutralization of weak acid by a strong base (cf. fig. 72). The sensitiveness of an indicator, and

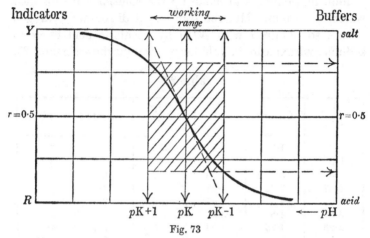

Fig. 73

the 'resistance' of a buffer, are both measured by the slope of the curve, $dr/d\,(p\mathrm{H})$, where $r = \mathrm{acid}/(\mathrm{acid} + \mathrm{salt})$ or $\mathrm{R}/(\mathrm{R} + \mathrm{Y})$, and it is seen that the slope is a maximum at the mid-point, where $p\mathrm{H} = pK$ and $r = 0.5$. All indicators and buffer systems give replicas of this curve, the position being always defined by the fact that the mid-point is given by pK.

The preparation of buffer solutions *

Citrate buffers (Sörensen).

In respect of ease of preparation, range of $p\mathrm{H}$ covered (1–7) and insensitiveness to changes of temperature, these buffer mixtures are undoubtedly outstanding, but in common with all buffers based on polybasic acids (see also phosphate buffers below) they have the disadvantage of varying ionic strength (or salt content). In experiments where constancy of ionic strength is important, adjustments may be made by the addition of calculated amounts of potassium or sodium

* Britton, *Hydrogen Ions* (Chapman and Hall, 1942), vol. I, chap. XVI.

chloride (some indicators are rather sensitive to change of ionic strength, see below, p. 236).

Dissolve 21·00 g. of pure citric acid in 200 ml. of normal sodium hydroxide, and make up the solution with distilled water to 1000 ml. Mix the solution of di-sodium hydrogen citrate so obtained with $N/10$ hydrochloric acid or $N/10$ sodium hydroxide in the volume proportions shown in table 23.

Table 23

$N/10$ HCl	Citrate	pH	$N/10$ HCl	Citrate	pH
9·0	1·0	1·17	5·0	5·0	3·69
8·0	2·0	1·42	4·5	5·5	3·95
7·0	3·0	1·93	4·0	6·0	4·16
6·67	3·33	2·27	3·0	7·0	4·45
6·0	4·0	2·97	2·0	8·0	4·65
5·5	4·5	3·36	1·0	9·0	4·83
5·25	4·75	3·53	0·5	9·5	4·89
			0·0	10·0	4·96

The pH values may be taken as correct between 10 and 70° C.

$N/10$ NaOH	Citrate	pH$_{18°}$	$N/10$ NaOH	Citrate	pH$_{18°}$
0·0	10·0	4·96	3·0	7·0	5·57
0·5	9·5	5·02	4·0	6·0	5·98
1·0	9·0	5·11	4·5	5·5	6·34
2·0	8·0	5·31	4·75	5·25	6·69

The pH values increase with rise of temperature (0·04/10° between 10° and 70°).

Acetate buffers.

These buffers may be prepared by the addition of standard acetic acid ($N/5$) to a fixed volume of standard ($N/5$) sodium hydroxide, a procedure ensuring constancy of salt content. Put into a graduated cylinder 10 ml. of $N/5$ sodium hydroxide, add the volume (ml.) of $N/5$ acid shown in table 24, and make up all the mixtures to 70 ml. with distilled water.

Table 24

| N/5 acid ... | 55·5 | 37·7 | 27·0 | 20·4 | 16·67 | 14·2 | 12·65 | 11·7 | 11·0 |
| pH   ... | 4·0 | 4·2 | 4·4 | 4·6 | 4·8 | 5·0 | 5·2 | 5·4 | 5·6 |

The pH values may be taken as correct between 10° and 40°.

Phosphate buffers:

(A) Dissolve 9·08 g. ($M/15$) of KH_2PO_4 in 1 l. of water.

(B) Dissolve 11·87 g. of $Na_2HPO_4.2H_2O$ or 23·90 g. of $Na_2HPO_4.12H_2O$ ($M/15$) in 1 l. of water. Mix the solutions in the proportions by volume given in table 25.

Table 25

A (H_2PO_4)	B (HPO_4)	pH	A (H_2PO_4)	B (HPO_4)	pH
9·75	0·25	5·29	5·0	5·0	6·81
9·5	0·5	5·59	4·0	6·0	6·98
9·0	1·0	5·91	3·0	7·0	7·17
8·0	2·0	6·24	2·0	8·0	7·38
7·0	3·0	6·47	1·0	9·0	7·73
6·0	4·0	6·64	0·5	9·5	8·04

These pH values are correct between 10 and 70° C.

Phosphate buffers may also be made up from KH_2PO_4 and NaOH (free from carbonate). Mix 50 ml. of a solution containing 13·6 g. of KH_2PO_4 per litre ($M/10$) with the volume of $N/5$ sodium hydroxide shown in the table.

Table 26

| N/5 NaOH ... | 5·0 | 7·5 | 10·0 | 12·5 | 15·0 | 17·5 | 20·0 | 22·5 |
| pH$_{25°}$ ... | 6·21 | 6·44 | 6·64 | 6·81 | 6·98 | 7·17 | 7·40 | 7·75 |

Between 12° and 75° the total decrease of pH is 0·02.

Borate buffers.

Dissolve 12·40 g. of boric acid in 100 ml. of normal sodium hydroxide, and make up with distilled water to 1 l. Mix the solution with $N/10$ hydrochloric acid or $N/10$ sodium hydroxide, as shown in table 27.

Table 27

Na borate	$N/10$ HCl	$pH_{18°}$	$N/10$ NaOH	$pH_{18°}$
10·0	0·0	9·24	—	—
9·0	1·0	9·09	1·0	9·36
8·0	2·0	8·91	2·0	9·50
7·5	2·5	8·80	—	—
7·0	3·0	8·68	3·0	9·68
6·5	3·5	8·51	—	—
6·0	4·0	8·29	4·0	9·97
5·75	4·25	8·14	—	—
5·5	4·5	7·94	—	—
5·25	4·75	7·62	—	—

Decrease of pH with rise of temperature changes steadily from 0·03/10° at 7·62 to 0·13/10° at 9·97.

(1) *Determination of the working range and constant of an indicator*

(a) *Two-colour indicators (e.g. methyl red or bromthymol blue).*

To 25 ml. of distilled water add 5–10 drops of the indicator solution, and well mix. Put about 10 ml. of this solution into each of two test-tubes, and add one drop of reagent sodium hydroxide and one drop of dilute hydrochloric acid to the tubes respectively. Prepare in test-tubes approximately 10 ml. portions of a set of buffers to cover the working range of the indicator (citrate for methyl red, phosphate for bromthymol blue), and add to each about twice the amount of indicator solution used in making the first solution. Hold the two test-tubes with free alkali and acid one behind the other, so that by looking through both the tint for R = Y is seen. The pH of the buffer giving the nearest match to this tint gives pK.

(b) *One-colour indicators (e.g. phenolphthalein, p-nitrophenol).*

Prepare a solution of the indicator containing about 0·1 g./l. Choose five test-tubes of closely similar cross-section. Place in four of them 10 ml. specimens of a suitable set of buffers to

Table 28. *Working range and constants of some important indicators*

(a) *One-colour indicators*

Indicator	Range of pH	$pK_{25°}$	Solvent, concentration, etc.
2-6-Dinitrophenol	2·0– 4·0	3·65	0·03 % in water
2-4-Dinitrophenol	2·6– 4·4	4·00	0·05 % in water
2-5-Dinitrophenol	4·0– 5·8	5·11	0·05 % in water
p-Nitrophenol	5·6– 7·6	7·18	0·1 % in water
m-Nitrophenol	6·8– 8·6	8·26	0·3 % in water
Phenolphthalein	8·3–10·0	9·6	0·04 g. in 30 ml. alcohol + 70 ml. water (fades in conc. alkali)
Thymolphthalein	9·3–10·5	—	0·04 g. in 30 ml. alcohol + 70 ml. water (fades in mod. conc. alkali)

(b) *Two-colour indicators*

Indicator	Range of pH	pK	Solvent, concentration, etc.
Methyl orange	3·1–4·4	3·6	0·1 g. in 300 ml. alcohol + 200 ml. water (short spectral range)
Bromphenol blue	3·0–4·6	4·0	0·04 % in alcohol (long spectral range)
Bromcresol green	3·8–5·4	4·7	0·04 % in alcohol
Methyl red	4·3–6·3	5·0	As for methyl orange
Bromcresol purple	5·2–6·8	6·3	0·04 % in alcohol (very long spectral range)
Bromthymol blue	6·0–7·6	7·1	0·04 % in alcohol
Phenol red	6·8–8·4	7·8	0·02 % in alcohol (short spectral range)
Thymol blue	8·0–9·6	8·9	0·04 % in alcohol

cover the working range (borate for phenolphthalein, phosphate for nitrophenol) and add 2 ml. of the indicator solution to each. To the fifth tube add one drop of reagent sodium hydroxide, 1 ml. of the indicator solution and make up to

12 ml. with distilled water. The coloured buffer matching this last tube gives $pH = pK$.

(2) *The 'salt effect' on buffers and indicators*

If an acetate buffer (p. 232) is made up so that the actual concentration of the salt is normal, and is then diluted to ten times its bulk with water, the effective concentration (or *activity*) of the salt falls from about 0·66 to about 0·08 (see table 19, p. 177, for sodium chloride), that is, to about one-eighth of its original value, while the concentration of the unionized acid falls to one-tenth of its value. The buffer ratio acid/salt is thus somewhat decreased, and the corresponding pH increased (equation (3), p. 230), on dilution. If in the now diluted buffer sodium chloride is dissolved to give again a total normal salt concentration, the pH will decrease to its former value, although the actual concentration of the buffer salt has not itself been changed. With buffers in which both acid and salt are ions (phosphate, borate) this salt effect is less easily predictable, but is usually a decrease in pH (table 29). In so far as at least one of the constituents R or Y is an ion a salt effect will also be expected for indicators, and will lead to a shift of the colour. In ordinary work salt effects may be neglected when the *total* ionic concentration does not exceed $N/10$.

The aggregated salt effects upon buffer and indicator may be examined as follows: Prepare in test-tubes two sets of buffers covering the range of the indicator; to each of one set add an equal volume of $2N$ sodium chloride, to bring the final total ionic concentration to about normal. Compare the colours of the indicator in each corresponding pair of tubes. As examples may be taken (a) methyl red or bromcresol purple, in citrate-NaOH buffers (table 23, p. 232); here the effect is a very marked decrease in the *apparent* pH as indicated by the colour; (b) bromcresol green, in acetate buffers (table 24, p. 233); here little or no colour difference will be discernible, the effect on the buffer being compensated by an opposite effect on the indicator.

(The separate effects on buffer and indicator may be disentangled by determining the pH of the buffers by the electrometric method, exp. c, p. 223.)

The illustrative data shown in table 29 are taken from the work of Kolthoff.*

Table 29

Buffer type	pH	pH with $0.5N$ KCl	Apparent pH from indicator tint	Indicator
Phosphate	5·6	5·2	5·45	Bromcresol purple
	6·2	5·8	5·7	Methyl red
	6·2	5·8	5·85	p-Nitrophenol
	7·6	7·2	7·35	Phenol red
Borate	8·6	8·33	8·5	Phenolphthalein

(3) *To compare the dissociation constants (strengths) of two monobasic weak acids*

A set of buffers of the one acid is prepared by partial neutralization with sodium hydroxide, and a few drops of a suitable indicator are added to each. Specimens of the second acid are then titrated with sodium hydroxide in the presence of the indicator until colour matches with the first series are obtained. We then have from equation (3), p. 230

$$(H^+) = K_1 r_1 = K_2 r_2, \quad \text{or} \quad r_2/r_1 = K_1/K_2,$$

where $r = $ acid/salt, and K is the dissociation constant.

As examples may be taken acetic and lactic acids, which are required in about $N/5$ concentration. Titrate samples of each acid with $N/10$ sodium hydroxide, using phenolphthalein. Prepare in a set of four test-tubes (labelled S 1, 2, 4 and 6 respectively) mixtures each containing 10 ml. of the weaker (acetic) acid, and 1, 2, 4 and 6 ml. of $N/10$ sodium hydroxide respectively. Add to each tube a suitable indicator. Bromphenol blue or bromcresol green is very satisfactory for the

* *Rec. Trav. chim.* **41**, 54 (1922).

acids mentioned; methyl orange is less suitable owing to its narrow spectral range. Label a second set of tubes T 1, 2, 4 and 6 and put into each 5 ml. of the approximately $N/5$ solution of the stronger (lactic) acid. Add indicator and titrate the acid in each tube in turn with $N/10$ sodium hydroxide until a colour match is obtained with the tube of the same number in the S (acetic acid) series. (With bromphenol blue or methyl orange as indicator, tubes 6, and with bromcresol green, tubes 1, may be omitted.)

Example

10 ml. of acetic acid required 20·0 ml. of $N/10$ NaOH.
10 ml. of lactic acid required 18·2 ml. of $N/10$ NaOH.
1 ml. of alkali solution = 0·5 ml. acetic acid, and 0·55 ml. of lactic acid.

Bromcresol green as buffer indicator

Tubes T (no.)	$N/10$ NaOH for colour match	with	Tubes S (no.)	$N/10$ NaOH in tubes S
2	4·4 ml.		2	2 (ml.)
4	6·1		4	4
6	7·1		6	6
(each 5 ml. acid)			(each 10 ml. acid)	
A ml. lactic acid neutralized by alkali (column 2 above × 0·55)	B lactic acid remaining 5 − A	r_1 (lactic) B/A	r_2 (acetic)	r_2/r_1
2·42	2·58	1·07	9·00	8·4
3·35	1·65	0·49	4·00	8·2
3·90	1·10	0·28	2·33	8·3

Hence $\dfrac{K_{\text{lactic}}}{K_{\text{acetic}}} = 8 \cdot 3$ (mean).

(Values in the literature are $K_{\text{acetic}} = 1 \cdot 8 \times 10^{-5}$, $K_{\text{lactic}} = 1 \cdot 5 \times 10^{-4}$.

Ratio $\dfrac{\text{lactic}}{\text{acetic}} = 8 \cdot 3$.)

(4) *The determination of pH by the use of indicators*

Two methods are in common use: (*a*) depending on colour comparison with indicators in sets of standard buffers, and most easily operated with two-coloured indicators; (*b*) the titration method of Michaelis, which requires one-coloured indicators, but no buffers.*

(*a*) As an example of the method the pH of a $N/10$ solution of ammonium chloride, and hence the degree of hydrolysis, may be determined.†

Dissolve 1·07 g. of pure ammonium chloride (*Analar* brand) in *conductance* water (for preparation, see p. 185) to make 200 ml. of solution, which is kept from contact with air as far as possible. Prepare in test-tubes 10 ml. specimens of citrate-NaOH buffers pH 4·96–5·57 (table 23, p. 232), and add the same amount of methyl red to each tube (the amount of indicator should be such that the colour differences between the tubes are easily discerned), and to a (corked) tube containing 10 ml. of the ammonium chloride solution. Match the colour in this last tube with one of the buffer series, or if a match cannot be exactly found put it between two of the buffer series so that a gradation of colour is seen. If necessary, buffers with a narrower pH step may then be prepared, their pH being calculated by assuming a linear relation between sodium hydroxide added and pH over the narrow range concerned.

Since the degree of hydrolysis x is very small, the equations connecting [H$^+$], x and the basic dissociation constant K_b may be written in the simplified form

$$[\mathrm{H}^+] = \frac{x}{V} = \sqrt{\left(K_w/K_b \frac{1}{V} \right)}$$

(V = the dilution in litres). Hence calculate x and K_b, taking $K_w = 10^{-14}$.

The increase of hydrolysis on rise of temperature may be

---

* For a survey of methods consult Britton, *op. cit.* on p. 231.

† Cf. Hill, *J. Chem. Soc.* **89**, 1273 (1906).

shown as follows. Place in a test-tube, which is then securely corked, 10 ml. of the $N/10$ solution of ammonium chloride in conductance water, with a few drops of a solution of bromcresol green. On heating the contents of the tube by immersing it in a beaker of hot water, the original blue colour of the indicator changes to green, showing a decrease of pH. On cooling the blue colour is restored.

(b) *The indicator titration method.** As an example of the application of this method the pH of carbonic acid-bicarbonate buffers may be determined, and hence the dissociation constant of carbonic acid.

The following solutions will be required:

(1) $M/5$ sodium bicarbonate, 16·8 g./l.

(2) $M/25$ hydrochloric acid, and $N/10$ sodium hydroxide.

(3) Indicator: p-nitrophenol, 1·0 g./l.

On two glass-stoppered bottles each of 150–200 ml. capacity mark the 100 ml. level with adhesive paper. Place in one bottle 20 ml. of the bicarbonate solution (1), 5 ml. of the hydrochloric acid (2) (No. 1, table below), bring the total volume to the 100 ml. mark with distilled water free from carbon dioxide, and then add 1 ml. of the indicator solution (use a dropping tube). Replace the stopper, and mix the contents of the bottle with very gentle movement, avoiding violent shaking, leading to loss of carbon dioxide to the airspace above the liquid.

Dilute 20 ml. of the indicator solution with distilled water to 200 ml. and fill a burette with the diluted solution. Place in the second glass bottle 10 ml. of approximately $N/10$ sodium hydroxide and about 80 ml. distilled water. Titrate the alkaline liquid with the diluted indicator until the depth of yellow colour matches that in the bottle of carbonate buffer. The liquid in the titration bottle will be nearly but probably not exactly at the 100 ml. mark. Repeat the titration, adjusting the water added so that the final volume at which

* Michaelis and Gyemant, *Biochem. Z.* **109**, 165 (1920).

the colours are matched differs as little as possible from
100 ml.

Prepare the two other buffer mixtures shown in the table,
and proceed with them in the same way.

	1	2	3
NaHCO$_3$ $M/5$	20	20	10 (ml.)
HCl $M/25$	5	10	20 (ml.)

(It may be noted that the mixture 2 has the proportions of
the carbonate buffer in normal blood.)

The whole of the indicator dissolved in the x ml. of diluted
solution added to the strongly alkaline liquid in the titration
bottle must be in the yellow form; it follows that the amount
of indicator dissolved in $10 - x$ ml. of diluted solution must
remain in the carbonate buffer in the colourless form. From
the pK of the indicator (see exp. b, p. 234) the pH values of
the buffers are calculated from the equation

$$p\mathrm{H} = pK - \log\,[(10-x)/x].$$

Example

Indicator added to carbonate mixture (expressed as diluted solution) X (ml.)	Indicator for titration x (ml.)	$X - x$	$\dfrac{X-x}{x}$	$\log \dfrac{X-x}{x}$
(1) 10	7·5	2·5	0·33	$-0\cdot48$
(2) 10	5·6	4·4	0·79	$-0\cdot10$
(3) 10	2·0	8·0	4·00	$+0\cdot60$

$$p\mathrm{H} = pK - \log \frac{\text{colourless form}}{\text{yellow form}}$$

$$= pK - \log \frac{X-x}{x}$$

$$= 7\cdot18 + 0\cdot48 = 7\cdot66 \text{ (solution 1)}$$
$$+ 0\cdot10 = 7\cdot28 \text{ (solution 2)}$$
$$- 0\cdot60 = 6\cdot58 \text{ (solution 3)}.$$

The dissociation constant of carbonic acid

| | $N/25 = N/5$ acid | | $M/5$ NaHCO$_3$ added ml. | H$_2$CO$_3$ formed (as ml. of $N/5$ solution) | NaHCO$_3$ remaining (as ml. of $N/5$ solution) | $\dfrac{\text{H}_2\text{CO}_3}{\text{HCO}_3^-}$ (x) | log x |
	$N/5$ acid						
	Added ml.	Acid ml.					
(1)	5	1	20	1	19	0·053	− 1·275
(2)	10	2	20	2	18	0·112	− 0·950
(3)	20	4	10	4	6	0·667	− 0·175

$$pK = pH + \log x$$
$$= (1)\ 6\cdot38,$$
$$(2)\ 6\cdot33,$$
$$(3)\ 6\cdot40.$$

Mean 6·37.

$$K = 10^{-6\cdot37} = 10^{0\cdot63} \times 10^{-7} = 4\cdot3 \times 10^{-7}.$$

Chapter VII

VELOCITY OF CHEMICAL REACTION*

Velocity of reaction is measured by the number of gram-molecules of substance undergoing reaction in unit volume of the reacting system in unit time. The unit of volume chosen is almost always the litre, but the time unit may be a second, a minute, an hour, or even a day, depending on the magnitude of the velocity. In expressing a velocity of reaction it is therefore essential always to give the units explicitly, e.g. a velocity of 0·52 l./min. means that in 1 l. of the reaction mixture 0·52 g.-mol. of reagent undergoes chemical change in 1 min. When the unit of volume is the litre, the reaction velocity is expressed symbolically as $-dC/dt$, the negative sign being required when C refers to a reactant, for then dC is negative.

In a reaction between gases A and B according to the equation $A + B \to C + \text{etc.}$, reaction can only take place when molecules of A, B collide in pairs. Without making any assumption about the fraction of the total collisions A, B in which chemical action ensues, we may yet infer that the number of pairs A, B changing in unit time and in unit volume will be *proportional* to the total number of collisions taking place in unit time and volume, i.e. to Z, the collision frequency A, B. The kinetic theory shows that Z is proportional in turn to the product of the concentration C_A and C_B (for the actual value of Z, see p. 255). Hence

$$-\frac{dC_{A \text{ or } B}}{dt} = kC_A C_B. \tag{1}$$

The constant k is called the (reaction) velocity constant, and is numerically equal to the rate at which A and B must be supplied to the reacting system to maintain $C_A C_B = 1$.

* General references: Hinshelwood, *The Kinetics of Chemical Change* (Oxford Univ. Press, 1940); Moelwyn-Hughes, *Kinetics of Reactions in Solution* (Oxford Univ. Press, 1947).

Since relatively satisfactory methods have been found of calculating collision frequency Z in solution* it has become clear that a relation of the form (1) should apply to solutions. Experiment has, indeed, established its widespread validity for reactions both in gases and solutions, but for some reactions in solution k must at present be regarded as empirical, for its value determined by experiment may differ greatly from that calculable from a simple kinetic theory.

In general we see that the right-hand member of (1) will consist of a product of a number n of concentrations C. The number n of such terms required *to reproduce the experimental results* of velocity determinations is termed the *kinetic order* of the reaction. Experiment shows that we must distinguish between this order and the molecularity of the reaction, which is the number of molecules shown as reactants in the stoichiometrical equation. An apparent distinction may arise merely from the conditions of experiment, as the following chart shows:

Reaction $A + B \rightarrow C +$ *etc.* *Molecularity* $= 2$ (*bimolecular*)

Conditions of reaction	Kinetic order
(1) A and B in comparable concentration (e.g. equivalent)	2
(2) A (or B) in large excess (as in exp. *a*)	1
(3) A catalyst surface is present, strongly adsorbing A (or B)	(frequently) 0 i.e. $-\dfrac{dC}{dt} = kC^0 = $ constant (see exp. *b*, p. 292)

A distinction may also be required for more fundamental reasons, arising out of the inherent mechanism of the reaction.

---

* Consult Moelwyn-Hughes, *Kinetics of Reaction in Solution*, pp. 8 seq. (Oxford Univ. Press, 1947).

Thus the reaction $2HI + H_2O_2 \rightarrow I_2 + 2H_2O$ (exp. 2) is found by experiment to have an order 2, and not 3, as the equation would suggest. Further examination has indicated that the actual chemical changes proceed in at least two stages:

(1) $HI + H_2O_2 \rightarrow HIO + H_2O$,   (2) $HI + HIO \rightarrow I_2 + H_2O$.

The rate of the first stage is relatively slow, but the second is almost instantaneous. The order is obviously that of the slower stage.

The maximum order shown *experimentally* may be termed the *basic* (or total) order. Research has shown that reactions are in general more complicated in mechanism than a single chemical equation would suggest, and there is a general tendency for reactions to show basic kinetic orders less than their molecularity; in fact it appears that reactions with an order higher than 2 are rare. Catalytic reactions in general exhibit orders less than their true molecularity (see (3), p. 261).

(1) The experimental determination of reaction velocity

The reaction to be first examined is the oxidation, in neutral solution, of iodide by persulphate:*

$$2I^- + S_2O_8^= \rightarrow 2SO_4^= + I_2.$$

The following solutions will be required:

(1) Potassium persulphate: dissolve 2·4 g. of the pure salt per 100 ml. of water, to give a solution of concentration approximately $0·18 N$.

(2) A solution of potassium iodide containing 6·6 g. of the salt per 100 ml. $(0·4 N)$.

(3) Standard sodium thiosulphate (about $N/10$).

(*a*) Dilute 40 ml. of the persulphate solution (1) to 200 ml. in a graduated flask, and then, with the aid of a pipette, put 50 ml. of the diluted solution $(0·036 N)$ into a conical flask,

---

* T. Slater Price, *Z. physikal. Chem.* **27**, 474 (1898); van Kiss and van Zombory, *Rec. Trav. chim.* **46**, 225 (1927).

of 150–200 ml. capacity, which is provided with a lead sinker, and placed in a thermostat regulated to about 25°. Put a second flask, with sinker, containing 50 ml. of the potassium iodide solution (2) in the thermostat beside the first (fig. 74).

Mix in a glass-stoppered bottle the above quantities (50 ml.) of iodide and diluted persulphate, close the stopper (which should be dry to allow the escape of air on warming) and immerse in a water-bath previously brought to about 50°.

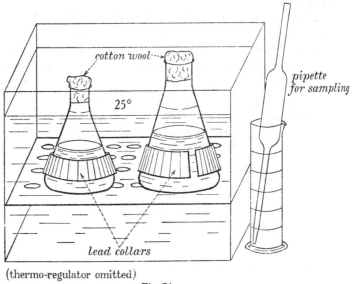

cotton wool

25°

pipette for sampling

lead collars

(thermo-regulator omitted)

Fig. 74

Leave without further heating until the conclusion of the velocity experiments.

Prepare from the $N/10$ solution of thiosulphate a $N/100$ solution by diluting 20 ml. to 200 ml. on a graduated flask. This solution should not be made until the time of the experiments, as it does not keep long unchanged in this very dilute concentration. Charge a burette with the diluted thiosulphate solution, and set beside the thermostat two large conical flasks (capacity about 300 ml.) each containing about 200 ml. of cold water.

After the above preparations have been finished the solutions in the thermostat should have reached temperature equilibrium (20 min.). If this is the case, pour the iodide into the persulphate (*not* the reverse way), and note on a watch or clock the time of mixing. Immediately after mixing agitate the flask to render its contents homogeneous. A short time before the expiry of an interval of 3 min. after mixing draw into a (*clean*) 10 ml. pipette a sample of the reaction mixture; adjust the level over the reaction flask, so that only 10 ml. is finally withdrawn from it. Discharge the pipette at the end of 3 min. interval into one of the conical flasks containing water. (In velocity experiments do not, as a rule, withdraw samples at arbitrary time intervals, the exact reading of which will probably cause confusion in an experiment in which orderliness is essential for success.) Titrate the sample with the thiosulphate already in the burette (starch should be used). The titration should be completed in not more than 3 min., so that a second sample can be taken after an interval of 4 min. from the first. All samples should be discharged into a considerable bulk of cold water, in order, by the large dilution, effectively to stop the reaction.

The period between the extraction of samples should be increased as the reaction rate diminishes (see example). At 25° the reaction is complete, i.e. all the persulphate is reduced, in about 1 hr. from the time of mixing.

After these experiments have been brought to an end, cool the stoppered bottle with tap water, open and titrate 10 or 20 ml. with the $N/100$ thiosulphate. This titration must be made, even if time has not allowed the completion of the main experiments, as it gives the 'iodine value' of the total persulphate used in the main reaction.

In working up the results according to the suggestions below, the concentration C of the persulphate may be taken as the titre (in $N/100$ thiosulphate) in column 4 below: it is unnecessary to calculate the absolute concentrations.

Example

Time intervals min. (1)	Time *elapsed* from mixing min. Δt (2)	Titrations with $N/100$ thiosulphate (10 ml. samples) (3)	Titre of persulphate, in ml. $N/100$ thiosulphate ($17 \cdot 7$ − column 3) (4)
0	0	0	17·7
3	3	3·45	14·25
4	7	7·25	10·45
3	10	8·90	8·80
5	15	11·65	6·05
5	20	13·60	4·10
10	30	15·80	1·90
12	42	17·20	0·50
18	60	17·60	(0·10)
	'Infinite' (from stoppered bottle)	17·70	

The experimental data may now be treated (i) directly, without first proceeding to a mathematical solution of equation (1), p. 243, (ii) and (iii) by methods dependent on solving the differential equation.

(i) Plot the concentrations of persulphate C (see note above) against the time elapsed, and draw a smooth curve through the points. 'Tangent' the curve at integral values of C (e.g. 16·0, 14·0, 12·0, etc.) by the method described on p. 9. Divide the values of $-dC/dt$ so found by the corresponding concentration C, and so evaluate the velocity constant; or plot dC/dt against C and find the slope of the (straight) line.

(ii) Since the initial concentration ($0 \cdot 2\,N$) of the iodide (A) in the reaction mixture was more than ten times that of the persulphate (B), C_A remains effectively constant, and equation (1), p. 243 becomes

$$-\frac{dC_B}{dt} = k'C_B \text{ (kinetic order} = 1, \text{ case 2, p. 244, above).}$$

The solution of this is

$$\ln C = \ln C_0 - k'(\Delta t), \qquad (2)$$

where Δt is the time elapsing for C_0 to fall to C. In decadic (ordinary) logarithms this is $\log C = \log C_0 - 0 \cdot 434\, k' \Delta t$. Plot $\log C$ against Δt, draw the 'best' straight line through the points, and find k' from its slope.

(iii) It is seen from equation (2) above that the period (of half-change) required for a given concentration C to fall to half its value is $\Delta t_{\frac{1}{2}} = \log 2 / 0 \cdot 434 k'$, and is independent of C. To use this relation (characteristic of first-order reactions) plot C against Δt as in (i), and read off times of half-change, taking various values of C.

(The mean value of k' obtained by the application of the above methods to the data given in the example is $0 \cdot 0728$ min.$^{-1}$.)

(b) Information about the effect of variation of iodide concentration upon the reaction velocity (which cannot be derived from the foregoing experiments) might in theory be obtained by reversing the procedure, viz. by taking an initial excess of persulphate, but practical difficulties arise owing to the precipitation and volatilization of iodine, when the reaction mixture contains only a small concentration of iodide. Further experiments similar to the above may, however, be carried out, in which the initial concentration of iodide C_A^0, although still large compared with that of the persulphate, is varied. The velocity constants k' will then be proportional to C_A^0 or to its square, according as the basic order of the reaction is 2 or 3.

Repeat exp. 1, but take, in place of 50 ml. of the iodide solution, a mixture of 25 ml. of the iodide solution with 25 ml. of $0 \cdot 4 N$ potassium chloride ($29 \cdot 8$ g./l.). As the reaction velocity will be reduced, the time intervals may be lengthened (to about double). The 'infinity' titration need not be repeated.

(The rates of all reactions between ions are subject to a 'salt effect', i.e. they are markedly influenced by the total effective ionic concentration, or ionic strength. It is therefore necessary to compensate the loss of ionic strength when the amount of iodide is reduced, by the addition of another electrolyte, such as potassium chloride, which is not acted upon by the persulphate. A short account of the 'salt effect' will be found on p. 252.)

(c) *The saponification of phthalide by alkali.**

The reaction to be studied can be represented as

It will be noticed that phthalide is a γ-lactone. It can be readily prepared by the alkaline reduction of phthalimide.†
When the alkali salt is acidified the lactone is regenerated.

In this example the molecularity is 2. It will be found that the velocity experiments can be satisfactorily interpreted by the equation

$$-\frac{dC}{dt} = kC_A C_B, \tag{3}$$

and that therefore the basic order is also 2. If for simplification we mix the reactants at the beginning in equivalent proportions, then C_A and C_B will remain equal throughout the reaction, and (3) reduces to

$$-\frac{dC}{dt} = kC^2,$$

of which the solution is $1/C - 1/C_0 = k\,(\varDelta t)$, in which C_0 is the initial concentration of either A or B.

If the purity of the specimen of phthalide is in doubt, it should be recrystallized from aqueous alcohol as follows: Dissolve the lactone in the minimum amount of boiling rectified spirit, and to the solution add water until incipient crystallization is seen. Cool in ice-water, when the lactone should be obtained colourless and melting sharply at 73°.

---

* Tasman, *Rec. Trav. chim.* **46**, 660 (1927).

† See *Organic Syntheses*, **16**, 71 (J. Wiley and Sons).

Weigh accurately 0·67 g. of phthalide ($M = 134$), and dissolve it in 225 ml. of distilled water free from carbon dioxide (p. 139), contained in a 400–500 ml. flask, and warmed to about 40°. Place the preparation in the thermostat (25°), at the same time putting about 50 ml. of an $N/5$ solution of sodium hydroxide* in a small flask beside the first. Charge a burette with $N/100$ hydrochloric acid in readiness for titration. A supply of ice should be available to chill the samples and arrest the reaction. As soon as the two solutions have taken the temperature of the thermostat, add with a pipette 25 ml. (one equivalent) of the alkali to the phthalide solution and well mix by shaking. As soon as possible after mixing take out a 25 ml. sample of the reaction liquid, run the sample at once into about 100 ml. of ice-cold water free from carbon dioxide, and titrate with $N/100$ HCl, using phenolphthalein as indicator, and adding the acid until only the faintest pink colour remains. Proceed to remove, chill and titrate further samples at regular intervals as shown in the example below.

Example

0·64 g. of phthalide: added 25 ml. of 0·192 N NaOH. Titrations of 25 ml. samples with $N/100$ HCl. Temperature 25·5°.

Time min.	Titration ml. $N/100$ HCl (a)	$(1/a) \times 10$	$10/a - 10/a_0$ (b)	$b/\Delta t$
0	28·30	0·353	—	—
5	16·40	0·610	0·257	0·051
10	11·60	0·862	0·509	0·051
15	8·75	1·143	0·790	0·052
20	7·15	1·399	1·046	0·052
30	5·40	1·852	1·499	0·050
40	4·60	2·174	1·821	0·046

* To ensure the absence of carbonate the solution of sodium hydroxide should preferably be prepared direct from sodium, dissolved in alcohol. If the solution is not exactly $N/5$, the weight of phthalide required for equivalence should be calculated accordingly.

The influence of ionic strength upon the velocity of ionic reactions*

The ionic strength of a solution is defined as $\mu = \frac{1}{2}\Sigma c_i z_i^2$, and is thus half the sum of the products of the ionic concentrations c_i into the squares of the corresponding charges z_i.

Ionic strengths of molar solutions of typical electrolytes

Type of electrolyte	Ionic strength
Uni-univalent (HCl, KCl, NaNO$_3$, KOH)	$\frac{1}{2}(1 \times 1 + 1 \times 1) = 1$
Uni-divalent (Na$_2$SO$_4$, H$_2$SO$_4$, BaCl$_2$)	$\frac{1}{2}(2 \times 1 + 1 \times 4) = 3$
Di-divalent (MgSO$_4$)	$\frac{1}{2}(1 \times 4 + 1 \times 4) = 4$

It is now recognized as an experimental fact that when ions react in very dilute solution (μ not greater than 0·002), the velocity constant k bears an exponential relation to the ionic strength:

$$\ln k = \ln k_0 + z\sqrt{\mu}. \qquad (4)$$

It is clear that k_0 is the (ideal) velocity constant when the reaction occurs in infinitely dilute solution ($\mu = 0$). When the reacting ions are of like sign the factor z is positive, and when they have different signs z is negative. Change of ionic strength may thus cause either an increase or a decrease of reaction velocity.

Brönsted† first gave an explanation of the form of (4) as follows. A reaction, e.g. $A^- + B^- \rightarrow products$, is regarded as proceeding through a fugitive complex $X^= = A^-.B^-$, and thus occurs in the stages

$$A^- + B^- \leftrightharpoons X^=, \qquad (a)$$
$$X^= \rightarrow final\ products. \qquad (b)$$

---

* Consult for a full account Moelwyn-Hughes, *Kinetics of Reactions in Solution*, chap. IV (Oxford Univ. Press, 1947).

† *Z. physikal. Chem.* **102**, 169 (1922).

If K_a is the equilibrium constant of (a), we have

$$a_X = C_X f_X = K_a a_A a_B = K_a C_A C_B f_A f_B.$$

Here a represents the activity, and f the activity coefficient (see Chapter VI, pp. 178, 195). According to (b) the rate of production of the final products is proportional to C_X; hence the rate of reaction is given by

$$k' C_X = k' K_a C_A C_B \frac{f_A f_B}{f_X},$$

and the velocity constant, given by rate of reaction $= k C_A C_B$, is

$$k = k' K_a \frac{f_A f_B}{f_X} = k_0 F. \tag{5}$$

The modern theory of electrolytes (see pp. 176 sqq.) shows that in dilute aqueous solution, and for ions of low charge

$$-\log f = 0 \cdot 507 z^2 \sqrt{\mu}, \tag{6}$$

where z is the charge on the ion of coefficient f. Combining (5) and (6), and remembering that $z_X = z_A + z_B$, we find finally

$$\log k = \log k_0 + 1 \cdot 014 z_A z_B \sqrt{\mu}. \tag{7}$$

Thus the empirical factor z in (4) contains the product of charges $z_A z_B$, and its change of sign is simply explained.

The reaction between iodide and persulphate (exp. 1, p. 245) is considered to proceed in the stages

$$\mathrm{I^-} + \mathrm{S_2O_8^{=}} \leftrightarrows (\mathrm{S_2O_8 \cdot I})^{=}, \tag{a}$$
$$(\mathrm{S_2O_8 \cdot I})^{=} + \mathrm{I^-} \rightarrow 2\mathrm{SO_4^{=}} + \mathrm{I_2}. \tag{b}$$

The first stage (a) is the slow, and therefore the rate-determining, step. The basic order should thus be 2 and not 3. An investigation at sufficiently low concentrations* indicates agreement with the requirements of equation (4).

Owing to the sensitiveness of the activity coefficients of ions to change in ionic strength, it is easy to confirm Brönsted's theory by numerous examples of ionic reactions,† but there is no reason why the conception of the reaction (or transition)

---

* King and Jacobs, *J. Amer. Chem. Soc.* **53**, 1704 (1931).

† Meolwyn-Hughes, *loc. cit.* on p. 252.

complex should be limited to such special types of reaction. Brönsted himself adumbrated a more general application,* and more recently Polanyi, Eyring, and others have endeavoured to extend the scope of the idea, which falls naturally into place in the modern theory of activation (see below).

(2) The influence of change of temperature upon reaction velocity

In general, reaction velocity is extremely sensitive to change of temperature, and in the great majority of known reactions rises markedly with even a small increase of temperature. The increase is expressed by the ratio of the velocity constants, $k_{T_1}/k_{T_2} = G$, which is called the *temperature coefficient*. The relation, deduced from experiment, between G and the temperatures concerned is of the form

$$\ln G = B \left[1/T_2 - 1/T_1 \right], \qquad (8)$$

G being found to have values between about 2 and 4 for a temperature range of 10°. Such a rapid change cannot be attributed to the increase in collision frequency Z (see p. 243), which is not more than 2 % for a rise of 10°, but can be completely explained by assuming that chemical reactions occur only during collisions of exceptional violence, or in those in which the molecules are in an unusually active (highly energized) condition.

We owe to Arrhenius the first suggestions that have developed into the modern theory of activation. He pointed out that if we regard (8) as deriving from a more general equation

$$\ln k = A - B/T, \qquad (9)$$

where A and B are specific constants, independent of temperature, then we have

$$k = Qe^{-B/T}, \qquad (10)$$

where $\ln Q = A$.

* *Z. physikal. Chem.* **115**, 337 (1925).

Now an expression for k of this form would result from the following theory:

Suppose that a collision between A and B only leads to chemical reaction when it takes place with a minimum energy E_a (per g.-mol.), contributed by the two molecules in the portions E_A and E_B respectively; and that the Maxwell distribution of energy holds for the reacting gas mixture. Then the collision frequency Z_a of pairs of potentially reactive molecules is (in l./sec.)

$$Z_a = \left[\frac{\sigma_A + \sigma_B}{2}\right]^2 \sqrt{\left\{8\pi R T \left(\frac{1}{M_A} + \frac{1}{M_B}\right)\right\}} \, n_A \, n_B \, e^{-(E_A + E_B)/RT}.$$
(11)

σ = effective cross-section of molecules, M = molecular weight, n = number of molecules per litre. The variation with T of $\sqrt{T}$ may be neglected in comparison with the much greater variation of the exponential term.

The velocity constant is equal to the velocity of reaction when $n = 1$, and therefore

$$k = Z e^{-(E_A + E_B)/RT} = Z e^{-E_a/RT},$$
(12)

where Z is nearly independent of temperature.

Correlating (12) and (10) we have $B = E_a/R$, and $Q = Z$. Hence E_a may be calculated from experimental values of B, and for the great bulk of reactions turns out to lie between 10,000 and 50,000 cal. (per g.-mol.). The high energy per reacting pair of molecules $= E_a/N$ (N = Avogadro number) may be contrasted with the average energy

$$\left\{\tfrac{3}{2} RT/N\right\} = 3000/N \text{ cal.}$$

for a perfect monatomic gas at 1000° abs. It may be noted in particular that a temperature coefficient $(k_1/k_2)^{10°} \simeq 2$, round 100° C., requires $E_a \simeq 20,000$ cal.

Owing to the logarithmic form of equations (8) and (9), very accurate values of k are necessary to give even moderate accuracy in E_a. This inherent difficulty has impeded the experimental verification of the theory of activation, but modern research* has indicated that E_a, as calculated by the method given above, may be far from temperature independent, and therefore that at present the theory is incomplete.

The velocity constant and temperature coefficient of the reaction
$$2HI + H_2O_2 \rightarrow I_2 + 2H_2O \ \textit{(in acid solution)}$$

This reaction was studied in great detail by Harcourt and Esson,† whose pioneer work contributed much to the foundations of a theory of reaction velocity. They showed that the basic kinetic order is 2, and not 3, the observed velocity being accurately expressible as

$$-\frac{dC}{dt} = k\,[HI]\,.\,[H_2O_2]. \tag{13}$$

The probable explanation has already been suggested above (p. 245).

The following solutions will be required:

(1) A 2-vol. solution of H_2O_2 (0·6 % H_2O_2).

(2) Sulphuric acid, made by mixing 1 vol. of concentrated acid with 2 vol. of water.

(3) A standard solution of sodium thiosulphate (about $N/10$).

(4) Freshly prepared starch solution.

Dissolve 2 g. of potassium iodide in 500 ml. of distilled water contained in a large flask F (fig. 75). Add by means of a graduated cylinder 30 ml. of the sulphuric acid (2) and well mix with the iodide solution. Stand the flask, preferably on a rubber pad R, in a bath containing ice and water. Arrange a retort stand and ring G to keep the flask upright, but allow

---

* Moelwyn-Hughes, *Proc. Roy. Soc.* A, **164**, 295 (1938). Consult also on the theory of activation: (a) Hinshelwood, *Kinetics of Chemical Change*, Chapter 3 (Oxford Univ. Press, 1940); (b) Farrington Daniels, *Chemical Kinetics*, George Fisher Baker lectures (Cornell Univ. Press), 1938.

† *Phil. Trans.* A, **186**, 817 (1895).

sufficient freedom for the flask to be shaken without removal
from the bath. Shake the flask occasionally to promote the
cooling of its contents. Put into separate test-tubes with the
aid of pipettes, 20 ml. of the solution of hydrogen peroxide (1),
and 10 ml. of the starch solution; immerse the two tubes also
in the cooling bath. While the solutions are thus cooling,
titrate the solution of hydrogen
peroxide by the following method
(if it is only desired to determine
the temperature coefficient, and
not the absolute velocity constants, this titration will not be
required).

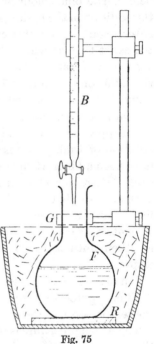
Fig. 75

Place 20 ml. of sulphuric acid
(2) diluted to 100 ml. with water
in a conical flask. Add and
dissolve about 2 g. of potassium
iodide (or add 20 ml. of a 10 %
solution of the iodide, free from
iodine, when making up the acid
to 100 ml.). Warm the acidified
iodide solution to about 30°, add
10 ml. of the solution of hydrogen
peroxide, mix by shaking and
then allow the whole to stand for
10 min. Finally, titrate the iodine
produced, with the standard thiosulphate (3).

The temperature of the solution in the flask F should now
be recorded; if it is found to be not above 2° the further
experiments may be started, and during their course the
temperature will remain practically unchanged at this initial
value.

Arrange a burette B (preferably, but not essentially, of small
pattern, with total capacity 10 ml.), so that it will deliver into
the large flask; fill the burette with thiosulphate solution (3).
If a stop-watch or clock is available, set it to its zero mark.

Pour into the flask F from the test-tubes, and well mix by shaking, first the starch, and then the 20 ml. of hydrogen peroxide. Start the stop-watch, or note the time, as the latter is added. Run from the burette (direct into the solution and *not down the neck of the flask*) 1 ml. of the $N/10$ thiosulphate, and well mix immediately with the contents by shaking as before. The blue colour of starch-iodine, which developed on the addition of the hydrogen peroxide, will now be discharged. After about 2 min. this colour will suddenly reappear, and the exact time at which this happens must be noted (but do *not* stop the watch). The colour is then again discharged by the addition of another 1 ml. of the thiosulphate, which is immediately well mixed by shaking, and the time of re-appearance again observed. Proceed in this way with timing the reappearance of the blue colour, and its discharge by the addition of 1 ml. portions of thiosulphate, until 10 ml. in all have been added. It should be realized that it is *the reappear-ance of the colour* which must be timed, and not the addition of the thiosulphate, which, however, should be added with as little delay as an accurate volume measurement will allow. The intervals between the appearances of the colour gradually lengthen, until at the end of the experiments more than 3 min. elapse.

Finally, again note the temperature of the solution, and if it differs appreciably from the initial value, assume a mean as the temperature of the experiments.

The whole procedure is repeated, in the same way, and with the same quantities and concentrations of solutions, except that the ice-water bath is replaced by one of tap water at room temperature, which is noted as before, and recorded.

(N.B. If room temperature is above 16° the intervals will be rather too short for accurate observation, unless a more con-centrated solution of thiosulphate is used. It is not desirable to increase the volume of the added portions above 1 ml., as it is assumed in the calculations that the reaction volume is effectively constant. At 20°, using $N/5$ thiosulphate in 1 ml. portions, the first (and least) interval is about 1 min.)

Example

Titration of H_2O_2—20 ml. required 28·10 ml. of thiosulphate (14·22 g. $Na_2S_2O_3$ per litre). 1 ml. portions of the above thiosulphate added to the reaction mixture.

Total thio-sulphate added ml.	Time intervals		Total time from beginning (sec.)		Ratio of total times
	At 2·0°	At 14·1°	At 2·0°	At 14·1°	
1	2 m. 14 s.	0 m. 51 s.	134	51	(2·63)
2	2 17	1 2	271	113	2·40
3	2 23	0 58	414	171	2·42
4	2 28	1 0	562	231	2·44
5	2 36	1 5	718	296	2·42
6	2 43	1 7	881	363	2·43
7	2 54	1 10	1055	433	2·44
8	3 3	1 14	1238	507	2·44
9	3 11	1 18	1429	585	2·45
10	3 30	1 27	1639	672	2·44
					Mean 2·43

The reaction is so conducted that the total volume of the reaction mixture is effectively constant, and the iodide is also at constant concentration, since it is always regenerated by the reaction

$$I_2 + 2Na_2S_2O_3 = Na_2S_4O_6 + 2NaI.$$

Hence the reaction between the peroxide and the iodide reduces to one of first order (see equation (13)). The constant ratios in the last column of the table give therefore the ratio $k_{14·1°}/k_{2·0°}$, since they correspond to the disappearance of equal fractions of the peroxide (equivalent to the thiosulphate added).

Calculate from equation (8) the temperature coefficient for the interval 10–20°. [First evaluate B, and then calculate G for $T_1 = 293°$, $T_2 = 283°$.] Record the activation energy

$$E_a = BR = 2B \text{ (cal.)}.$$

The velocity constants. These may be obtained by any of the methods (i), (ii) or (iii) on p. 248. The titre of the peroxide is found from the titration with thiosulphate, and the known portions of thiosulphate added during the reaction, e.g. for the example above it would be $28 \cdot 1 - n$, where n is the number of ml. of thiosulphate added. In using method (iii) take the period of 1/3 or 1/4 change.

(3) Catalysed reactions

These may be divided into two classes which, however, are not always easily distinguishable in practice:

(1) The catalyst is a substance present in the same physical condition and in the same phase as the other reactants, but the nature of the reactions is such that it finally emerges from the reactions in a chemically unchanged condition. This type is termed *homogeneous* catalysis.

(2) The catalysis occurs in a surface film of adsorbed reactants; this is *heterogeneous, contact,* or *surface* catalysis.

Only examples of the first type will be given here, as the second is more appropriately treated in the next chapter, on surface chemistry (pp. 290 sqq.).

There is no doubt that the homogeneous catalyst is to be regarded as a reactant, and the reaction as proceeding in stages, in the last of which the catalyst reappears. A typical and at the same time simple case is provided by one of the examples in the following experiments. The oxidation of iodide by ferric iron $2Fe^{+++} + 2I^- \rightarrow 2Fe^{++} + I_2$ proceeds relatively slowly (at room temperature), and ultimately comes to an equilibrium. If a cuprous salt is added, the reaction $Fe^{+++} + Cu^+ \rightarrow Fe^{++} + Cu^{++}$ occurs with great rapidity, and is immediately succeeded by the equally fast reaction

$$2Cu^{++} + 2I^- \rightarrow 2Cu^+ + I_2,$$

which regenerates the cuprous ion. It is just this last possibility that makes the cuprous ion an effective catalyst for the original reaction. Catalytic reactions thus provide further examples of reactions in which the molecularity, which should include the catalyst, is greater than the kinetic order.

It will be clear from the above explanation why catalysts for oxidation (examples (i) and (ii) below) are so often found among the ions and oxides of those metals that can exist in higher valency states unstable in the conditions of the reaction.

Autocatalysis arises when an effective catalyst is produced by the reaction itself, as in the bromination of acetone (catalyst HBr) and in the reaction of permanganate and oxalic acid (catalyst Mn^{++}).

Typical experiments illustrating (homogeneous) catalysis

(i) A portion of 50 ml. of $N/10$ potassium permanganate is acidified with 20 ml. of dilute sulphuric acid and mixed with an equal volume of an equivalent solution of oxalic acid at room temperature. The rate of interaction as estimated by removal of colour is such that reaction is only complete in half an hour or longer. The experiment is repeated with the previous addition of one drop of 10 % manganous sulphate solution to the permanganate. Decolorization is now complete in a few seconds.*

(ii) Place in each of two flasks 100 ml. of 2-volume hydrogen peroxide. To one (*A*) add 5 ml. of a 10 % solution of potassium iodide, and to the other (*B*) 5 ml. of water. At intervals of about 10 min. remove from flask *A* 10 ml. samples, and add each of them to a previously prepared mixture of 10 ml. dilute (20 %) sulphuric acid, 10 ml. of the 10 % solution of iodide used above, and about 1 ml. of a 10 % solution of ammonium molybdate. Titrate the liberated iodine with a standard solution of sodium thiosulphate (about $N/10$), and thus determine the amount of hydrogen peroxide in the sample. Titrate by the same method one or two samples from the control flask *B*. After a few minutes from the addition of the

---

* For the probable reactions, consult Bradley and van Praagh, *J. Chem. Soc.* 1938, p. 1624.

iodide the rapid decomposition of the peroxide in A will be demonstrated by the appearance, especially on shaking, of numerous bubbles of oxygen.

The catalysed reaction proceeds, as usual, in stages, of which the first is probably $H_2O_2 + I^- \rightarrow IO^- + H_2O$ (see p. 245 and exp. p. 256). In the absence of acid, I^- and IO^- do not interact, but the latter and H_2O_2 undergo mutual reaction of a type frequently exhibited by a pair of oxidizing agents:

$$IO^- + H_2O_2 \rightarrow I^- + H_2O + O_2,$$

the catalyst I^- being thus regenerated. This explanation is in agreement with the facts that the reaction is of the first order, and the velocity of decomposition of the peroxide is proportional to the concentration of the iodide.* The method of titration provides an example of oxidation catalysis, by molybdic acid, whose higher state is probably that of a permolybdic acid.

(iii) To each of two flasks A and B each containing 250 ml. of distilled water add 10 ml. of 10 % potassium iodide and 15 ml. of dilute hydrochloric acid. To flask A add 2 drops of 10 % solution of copper sulphate, and some freshly made starch solution to both. After 1 min. decolorize the contents of A by titrating with a few drops of $N/20$ thiosulphate, and then add with a pipette 10 ml. of the same thiosulphate solution to each flask. Finally, add to each flask at about the same time 5 ml. of 10 % ferric chloride solution and well mix by shaking. A blue colour appears in A after only a few minutes, but about half an hour elapses before a similar colour appears in B. The thiosulphate serves only to demonstrate the rate at which the iodine is being produced.

The mechanism of this reaction has already been suggested (p. 260).†

(iv) *Autocatalysis. The bromination of acetone.*‡ To 25 ml. of acetone contained in a boiling tube add from a burette 1 ml. of a 5 % solution of bromine in carbon tetrachloride. No de-

---

* For a full account, see Walton, *Z. physikal. Chem.* 47, 185 (1904).

† Cf. F. L. Hahn and Windisch, *Ber.* 56, 598 (1923); *Chem. Soc. Abstr.* 2 (1923), p. 262.

‡ Lapworth, *J. Chem. Soc.* 85, 30 (1904).

colorization is apparent. Warm in a water-bath to about 50°, when decolorization occurs suddenly. Cool again to room temperature with tap water, and then add another 1 ml. of the bromine solution; the colour is now seen to diminish with some speed at room temperature. On making further additions of bromine the speed of uptake steadily increases, until after about 5 ml. in all have been added, the tap of the burette may be fully opened and the colour disappears as fast as the bromine can be added, fumes of hydrobromic acid being simultaneously liberated:

$$CH_3COCH_3 + Br_2 + (HBr) \rightarrow CH_2Br.CO.CH_3 + 2H^+ + 2Br^-.$$

The mechanism of the reaction depends on the enolization of acetone, which is catalysed in proportion to $[H^+]$.

Catalysis by hydrogen ion

Probably no other catalytic agent exhibits such protean activity as the hydrogen ion, to which the formula H_3O^+ should be assigned when it is found in aqueous solution; but it finds its most important use as a catalyst in reactions involving water as a reactant or resultant, especially esterification and hydrolysis. The hydrolysis of a compound AB may be assumed to take place as follows:

$$AB + H_3O^+ \rightarrow AH + BOH + H^+,$$
$$H^+ + H_2O \rightarrow H_3O^+,$$

the velocity of the hydrolysis being given by

$$-\frac{d[AB]}{dt} = \frac{d[AH]}{dt} = \frac{d[BOH]}{dt} = k_s[H_3O^+][AB] = k'[AB],$$

since the value of $[H_3O^+]$ remains unchanged. The quantity $k'/[H_3O^+]$, equal to the constant k_s, is sometimes called the (hydrogen ion) *catalysis constant* or catalytic coefficient (Dawson), and is specific for a particular compound undergoing hydrolysis. Simple hydrolyses are therefore of the first kinetic order, but the apparent constant k' is proportional to the hydrogen-ion concentration (when this is not too high).

The practical treatment of many examples of catalysed hydrolysis is complicated by the participation of OH ion also as an active catalyst. This difficulty does not however arise in the hydrolysis of acetals and ketals, e.g.

$$CH_3CH(OC_2H_5)_2 + H_2O = CH_3CHO + 2C_2H_5OH.$$
(acetal)

$$(CH_3)_2C(OC_2H_5)_2 + H_2O = CH_3COCH_3 + 2C_2H_5OH.$$
(ketal)

These compounds are not appreciably hydrolysed by the catalytic agency of the OH ion, but are extremely sensitive to H_3O^+.*

(a) *The hydrolysis of acetal by hydrogen ion.*

In 200 ml. of distilled water, contained in a flask of Jena, Pyrex, or other 'resistance' glass, dissolve 5 g. of pure acetal ($M = 118$, $d_{20°} = 0.83$, b.p. 105°), thus obtaining an approximately $M/5$ solution. When the solution has assumed the temperature (25°) of the thermostat, and arrangements have been completed for titrating the samples (see method below), start the hydrolysis by the addition and mixing of 1 ml. of $N/10$ hydrochloric acid. The final concentration of H_3O^+ in the reaction mixture is thus $0.0005\ M$. Stopper the flask (loosely) to minimize the loss of aldehyde. Remove 20 ml. samples at approximately the intervals indicated in the example, and run each sample immediately into 20 ml. of a molar solution of sodium sulphite, as explained in the method of analysis.

Analysis of the samples. A suitable method of estimating the aldehyde† depends on the reactions

$$Na_2SO_3 + H_2O \rightleftharpoons NaHSO_3 + NaOH,$$

$$CH_3CHO + NaHSO_3 \rightarrow CH_3CH(OH)SO_3Na.$$

The first shows the ordinary hydrolytic equilibrium of sodium sulphite, and the second the formation of aldehyde-bisulphite compound, which, by the removal of bisulphite, disturbs the equilibrium, so that finally sodium hydroxide equivalent to the aldehyde is liberated. This alkali may be titrated with standard

* Skrabal and Mirtl, *Z. physikal. Chem.* **111**, 98 (1924); Brönsted and Wynne-Jones, *Trans. Faraday Soc.* **25**, 59 (1929).

† Cf. Sutton, *Volumetric Analysis*, p. 386, revised by A. D. Mitchell (Churchill, 1935).

acid. To ensure that the alkali titrated is actually equivalent to the aldehyde, a large excess of sulphite must be present.

Prepare a nearly saturated (molar) solution of sodium sulphite by dissolving 250 g. of the crystals ($Na_2SO_3 . 7H_2O$) in 1000 ml. of distilled water. To each of two portions of this solution, each of 20 ml., add 2 drops of phenolphthalein indicator, and sufficient $N/10$ hydrochloric acid to yield a pale pink colour (about 2·5 ml. of acid should be required). To one portion now add the sample withdrawn from the reaction mixture; after 1 min. titrate with $N/10$ hydrochloric acid, until the pale pink colours are again matched. Only the acid used in the *final* titration is to be taken as equivalent to the aldehyde in the sample. After titration reject the liquid from the second flask, but retain the other as a colour reference in the titration of subsequent samples. Calculate the acid required when the hydrolysis of the acetal is completed, from the weight of acetal originally taken, and so obtain the 'infinity' titration. It is not to be recommended to wait for completion in the reaction liquid itself, as the loss of aldehyde through volatilization will introduce serious errors.

Example

0·2248 M acetal and 0·0005 M hydrochloric acid. $N/10$ hydrochloric acid required for 20 ml. of reaction mixture, when added to the sulphite solution after the hydrolysis of the acetal is completed, is 0·2248/0·1 × 20 = 45·0 ml.

Time intervals min.	Titration ml. $N/10$ HCl	Titre of acetal (45 − column 2) ml. $N/10$ HCl
0	11·7	33·3
3	14·7	30·3
5	18·7	26·3
6	22·9	22·1
7	27·0	18·0
10	30·5	14·5
10	33·8	11·2
∞	(45·0)	0·0

$$k_{\log 10} = 1·22 \times 10^{-2}. \qquad k_s = \frac{1·22 \times 10^{-2} \times 2·30}{5 \times 10^{-4}} = 56.$$

Similar series of determinations for k, in which different amounts of catalyst, e.g. 1·5 and 0·5 ml. of $N/10$ hydrochloric acid, are used, will disclose that k is proportional to $[H_3O^+]$, and k_s is a constant.

(b) *The hydrolysis of diethyl ketal*, $(CH_3)_2C(OC_2H_5)_2$.

(For a convenient method of preparing this ketal, see below.)

The extreme sensitiveness of the ketals to hydrogen-ion catalysis, combined with the facts that the product acetone can be very exactly estimated, and is comparatively non-volatile in aqueous solution, render their hydrolysis peculiarly suitable for estimating low hydrogen-ion concentration, or pH (for other methods, see Chapter VI, pp. 223, 239).

To determine the catalysis constant k_s of diethyl ketal, the pH of the standard phosphate buffer* being assumed. To prepare the buffer dissolve 0·54 g. of pure KH_2PO_4 in about 100 ml. of distilled water. Add 20 ml. of $N/10$ sodium hydroxide, and make up to 200 ml. with water. The solution will now be 0·01 M in respect to both $H_2PO_4^-$ and $HPO_4^=$. Place the flask containing the prepared solution in the thermostat (25°). To start the reaction run into the solution from a burette, and dissolve, 2·5 ml. of pure ketal ($M = 132$, b.p. $= 114°$). Take 10 ml. samples at about the intervals indicated in the example, and run each of them immediately into a glass-stoppered bottle containing 2–3 ml. of concentrated sodium hydroxide solution (33 %). The hydrolysis is thus arrested. Add by a pipette 50 ml. of $N/10$ iodine solution. Stopper the bottle, and well shake for 1 min. Acidify with concentrated hydrochloric acid, and titrate the liberated iodine with $N/10$ sodium thiosulphate solution, and so estimate by the difference between the titration and 50 ml. the volume of iodine solution reacting with the acetone in the sample. Use freshly prepared starch as indicator. For the final 'infinity' sample use 60 ml. of $N/10$ iodine solution.

$$CH_3COCH_3 + 3I_2 + 4NaOH = CHI_3 + CH_3COONa + 3NaI + 3H_2O.$$

1 ml. $N/10$ iodine solution $= 0·00097$ g. acetone.

* See Chapter VI, Part 3, p. 233.

The comparatively low volatility of acetone in aqueous solution permits the infinity titration (complete hydrolysis) to be obtained directly.

Example

Buffer salts each $0.01M$. Temp. $25°$.

Time intervals min.	Titration ml. $N/10$ thio-sulphate	Titre of acetone ml. $N/10$ iodine
0	43·42	6·58
6	36·66	13·34
7	29·59	20·41
17	17·26	32·74
10	12·33	37·67
10	8·69	41·31
10	6·03	43·97
∞	9·27 (60 ml. iodine sol.)	50·73

$$k_{\log 10} = 1.30 \times 10^{-2}.$$

At $25°$ the pH of a mixture of KH_2PO_4 and Na_2HPO_4 each at concentration $0.01M$ is 6.96 (National Bureau of Standards Scale): $[H_3O^+] = 1.096 \times 10^{-7}$.

$$\text{Hence } k_s = \frac{1.30 \times 2.30 \times 10^{-2}}{1.096 \times 10^{-7}}$$

$$= 2.73 \times 10^5.$$

*The preparation of diethyl ketal**

Mix 1 mole (58 pts.) of acetone with 1 mole (160 pts.) of orthoformic ester and 3 moles (138 pts.) of absolute alcohol. Add 2–3 drops of concentrated sulphuric acid, and set the mixture aside for 24 hr. Then add a small quantity of aqueous ammonia until the liquid is alkaline, and a quantity of ether about equal in volume to that of the solution. Pour into ammoniacal water. Separate and wash the ether extract with a small quantity of water; dry it with potash, and then fractionate. Collect the fraction boiling 113–$115°$. The yield should be at least 70%.

* Hurd and Pollack, *J. Amer. C.S.* **60**, 1909 (1938).

Chapter VIII

SURFACE CHEMISTRY

(1) Surface energy and surface tension

Any molecule situated in the interior of a liquid is subjected to the attractive (e.g. van der Waals) forces of neighbouring molecules, exerted on the average equally in all directions. The same intermolecular forces must, however, exert a net inward attraction upon all molecules in or near the surface of the liquid (fig. 76). This tendency towards withdrawal from the

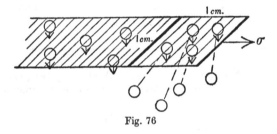

Fig. 76

surface leads obviously to the exposure by the liquid of as little surface as possible in given circumstances. Increase of surface demands transference of molecules to the surface (fig. 76), and therefore requires the input of work. The work necessary to increase a liquid surface by 1 cm.² isothermally is equal to the (specific) *surface free energy* (σ). If A is the area on the surface occupied by 1 g.-mol. of the liquid molecules, $A\sigma$ is equal to the *molecular surface free energy*. For molecules that may be regarded as spherical, $A = V^{\frac{2}{3}}$, where V is the molecular volume. The *relative surface free energy* of a liquid X is the ratio $\sigma_X/\sigma_{\text{water}}$. The surface free energy σ must be carefully distinguished from the surface total energy U_s; these two quantities are related by the thermodynamic equation

$$\sigma - U_s = T \frac{\partial \sigma}{\partial T}.$$

Near the freezing-point of a liquid U_s may be as much as three times σ.*

The dimensions of energy being given by the dimensional product force × length, we may *regard* the surface free energy as due to the necessity of overcoming a force F (force per unit length = dynes/cm.) acting *tangentially* to the surface. The work done against this force in creating 1 cm.² of surface is $F \times 1 = \sigma$ (fig. 76). The force F is called the surface tension, and is numerically and dimensionally equal to the surface free energy σ. Hence the terms surface free energy and surface tension are often used synonymously. An analogy, but with the work of opposite sign, may be seen in the expansion by 1 c.c. of a saturated vapour at pressure p; work $= p \times 1$.

Surface activity

The aqueous solutions of many substances, even when quite dilute, have surface free energies very much less than that of pure water. Substances producing such an effect are termed *surface-active* or *capillary-active*; the most familiar examples are the higher alcohols and fatty acids and their salts the soaps. The surface free energy of a weak (0·2 M) solution of amyl alcohol differs little from that of the pure alcohol (exp. *b*, p. 273), a fact suggesting that the dissolved alcohol is in much greater concentration at the surface than in the interior of the solution. The surface excess concentration or *adsorption* Γ can be predicted and calculated as a consequence of the lowering of the surface energy, by applying the principles of thermodynamics:†

$$\Gamma = -\frac{C}{RT}\left[\frac{\partial\sigma}{\partial C}\right]_T.$$

C = internal concentration of surface-active substance.

The surface energies of the *pure* alcohols and acids themselves differ little from those of the related paraffins. This fact

* Consult Lennard-Jones and Corner, *Trans. Faraday Soc.* **36**, 1156 (1940).

† Adam, *The Physics and Chemistry of Surfaces* (Oxford Univ. Press, 1941), p. 113.

can be explained by supposing that both in aqueous solution and in the pure liquids the surface molecules are oriented with the hydrocarbon chain outwards, and the OH or COOH group in the liquid (see also p. 279). It is in harmony with this view that the surface activity should rapidly increase with molecular weight, i.e. with the length of the carbon chain (exp. b, p. 273).

Since a surface-active substance lowers the surface energy it tends to spread spontaneously over the liquid surface, with a spreading force $F_s = \Delta\sigma$, where $\Delta\sigma$ is the lowering of the surface tension. A technique of measuring F_s directly has been extensively employed in recent years,* and it has been shown that the product $F_s \times A$, where A is the area occupied on the surface by 1 g.-mol. of the surface-active substance, is strikingly analogous to the product $p \times V$, which corresponds to three-dimensional changes.

The measurement of surface tension

The principal methods of determining surface tension, e.g. capillary rise, drop-weight, bubble pressure, etc., all utilize, directly or indirectly, the fact that liquid under a curved surface of radius r is subject to an excess or defect of pressure $2\sigma/r$ (dynes/cm.^{2}) according as r is positive or negative.

(i) *Capillary rise.*

This is the oldest, and remains the most direct, method of demonstrating and measuring surface tension. When a narrow glass tube (radius $d/2$) is dipped into a liquid which does not wet glass (e.g. mercury), the position and shape of the liquid meniscus in the tube is momentarily that of the curved broken line of radius r in fig. 77a. The excess downward pressure immediately forces liquid out of the tube to a level h, such that $hgD = 2\sigma/r$ (g = gravitational constant, D = density of the liquid). The radius of curvature r depends on the contact angle θ ($r = d/2 \cos\theta$). If the liquid wets the glass, the contact angle is zero, and the radius of curvature of the meniscus $= d/2$, the

* Adam, *op. cit.* on p. 269.

internal radius of the tube. The momentary condition, shown
by the broken line of fig. 77 b, is then one of pressure defect,

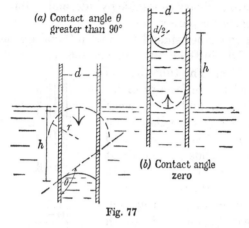

(a) Contact angle θ
greater than 90°

(b) Contact angle
zero

Fig. 77

and the outer liquid now forces liquid into the tube to a height
h, given by $hgD = 4\sigma/d$, and from this relation σ can be calculated when d is known.

(ii) *The drop-weight method* (fig. 78).

A drop of liquid *hangs* on a tube (external
diameter d) owing to the upward pressure
$2\sigma/r$, the weight of the drop just balancing
the force due to this pressure. If the contact
angle is zero, the radius of curvature $= d/2$, and

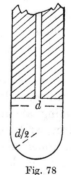

$$mg = \frac{4\sigma}{d} \times \frac{\pi d^2}{4} = \sigma \pi d. \qquad (1)$$

For reasons which will be given below,
this formula does not give accurately the
weight of the *fallen* drop.

Fig. 78

(a) *The method of capillary rise*

The difference Δh in rise in two tubes dipping into the
same liquid is

$$\Delta h = \sigma \left[\frac{4}{d_1} - \frac{4}{d_2} \right] \frac{1}{gD} = \text{constant } \sigma. \qquad (2)$$

Since to measure accurately the difference of height Δh is easier than to estimate the single height h, it is convenient to arrange two capillary tubes side by side, and calibrate them by observing Δh for pure water.

One of the tubes should have an internal diameter about 0·05 cm. (within the limits 0·06 and 0·04 cm.), and the other about 0·25 cm. The tubes should be selected for uniformity of internal diameter, and cut to such a length (about 25 cm.) that when standing in a 100 ml. graduated cylinder their ends project about 1 cm. above its top. Before use in an experiment the tubes must be cleaned by standing overnight in the graduated cylinder filled with acid dichromate. Immediately before they are required the tubes are removed, and the cylinder washed free from solution and then filled with distilled water. The tubes are thoroughly washed in running tap water, and then bound together by passing a small rubber band over them towards the ends; both are then immersed in the cylinder, their ends protruding about 1 cm. above the water, and the band being near the upper end. Dry the exposed end of each tube by touching with *clean* filter-paper, raise the tubes so that the meniscus in the finer tube comes well below the rubber band, and fix in this position with a retort clamp applied at the band (fig. 79). Move the cylinder up and down and ascertain that the meniscus in the fine tube follows these movements freely, and is not impeded by fine droplets

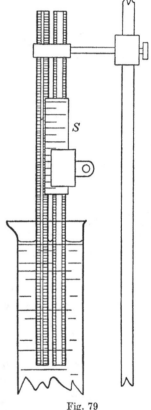

Fig. 79

in the upper part of the tube. If such are present the procedure should be repeated, more careful attention being given to the drying of the tops of the tubes. Add distilled water to the cylinder so that it is nearly filled. A short length of boxwood scale, S (about 7 cm.), carefully cut so that one end corresponds to a centimetre mark, is now attached to the tubes with the aid of a spring paper-clip, the centimetre mark being placed as exactly as possible against the meniscus in the broader tube: the difference of capillary height Δh is read off. The pair of tubes is now raised into a new position and another reading taken, after setting the scale as before. Continue the procedure until four or five readings of Δh have been made at different parts of the tubes.*

Assuming that the surface tension of water is 72 dynes/cm. (at room temperature = approx. 18°), calculate the constant of equation (2), taking a mean value of Δh.

(b) *The surface activity of the aliphatic alcohols*

Prepare $0 \cdot 2 M$ solutions by dissolving alcohols in the proportions $2 \cdot 3$ ml. of ethyl, $3 \cdot 6$ ml. of n-butyl, $4 \cdot 4$ ml. of amyl (free from pyridine) in 200 ml. of water.† Experiments with these solutions may be made along two lines, (1) to show that surface activity increases with chain length: for this, use the above solutions in turn, and in the order mentioned; (2) to plot a σ-concentration curve for any one alcohol. For this purpose amyl alcohol is the most suitable. Prepare 500 ml. of the $0 \cdot 2 M$ solution, and from it by dilution $0 \cdot 02$, $0 \cdot 05$ and $0 \cdot 1 M$ solutions (200 ml. of each). The experiments should proceed *from the more dilute to the stronger solutions*, and not in the converse way.

The graduated cylinder is filled with the alcohol solution, the tubes immersed and withdrawn several times to ensure their being filled with the solution, and then the operations of

---

* If a kathetometer is available, it will, of course, provide a more ready and accurate means of measuring Δh.

† If a burette is used to measure out the alcohol, its tap should not be greased.

drying the tops of the tubes and raising them conducted as in
the calibration (a). As the tubes are raised, solution should be
added so that the cylinder is always nearly full. Calculate σ
for the solutions from the observed mean $\varDelta h$ and the constant
obtained in (a): take $D=1$ for all solutions.

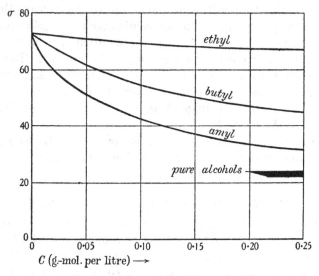

Fig. 80. Surface activity of lower alcohols.

The following are approximately the values which should be
obtained (fig. 80):

For 0·2 M solutions of	σ (dynes/cm.)
Ethyl alcohol	68
n-Butyl alcohol	47
Amyl alcohol	34

Amyl alcohol

Conc.	0·02	0·05	0·1	0·2 (*M*)
σ	60	52	43	34

(σ for the pure alcohol = 24 dynes/cm.)

(2) Interfacial tension

When two immiscible liquids (fig. 81) are in contact, a tension σ_{AB} analogous to surface tension is in general found at the interface. The surface tensions σ_A, σ_B are due to the (inward) attraction of molecules of A on surface molecules of A, and of molecules of B on surface molecules of B; surface tensions therefore measure attractions between *like* molecules. As the liquids are brought close together (fig. 81)

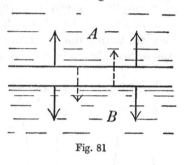

Fig. 81

and finally into contact, the cross-attractions of A and B come into play, and will act to reduce the separate tensions. Hence

$$\sigma_{AB} < \sigma_A + \sigma_B.$$

Thus the expression $\sigma_A + \sigma_B - \sigma_{AB}$ is an important and direct measure of the attraction between *unlike* molecules: it is usually called the *adhesional work* W_a, and is actually the work required (per cm.² of interface) to separate two immiscible liquids isothermally, in which process 1 cm.² of interface is destroyed, and 1 cm.² of new surfaces of A and B formed. The adhesional work may be contrasted with the *cohesional work* $W_c = 2\sigma_A$ or $2\sigma_B$, which is the work required (per cm.² of surface) to separate two *like* portions of liquids isothermally. It has been found that $\sigma_{AB} = \varDelta\sigma_{A,B}$ (Antonow's rule) is only approximately obeyed.

The drop weight method of comparing surface and interfacial tensions

As it is not usually possible to find the weight of a suspended drop, to which formula (1), p. 271, would apply, the method can only be applied to drops which have spontaneously separated from the dropping tube. The falling of a drop is a complicated process, portrayed in fig. 82.*

It will be seen that only a portion (60–70 %) of the drop falls, to be collected in a receiver beneath. However, experiments with liquids of known σ show that the ratio of the masses of drops m_A/m_B is given by σ_A/σ_B to about 5 %.

It is essential that (1) the dropping tip should be circular, of not less than 5 mm. external diameter, and ground plane at the orifice; this is best done with fine carborundum and tur-

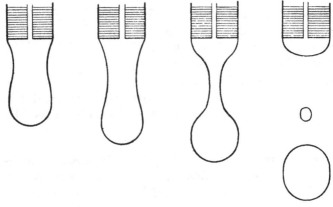

Fig. 82

pentine; (2) the drops form slowly, and fall by their own weight, unaided by the kinetic energy of liquid entering the drop from the tube. Within reasonable limits the weight of the drop is not influenced by the head of liquid in the delivery apparatus. Evaporation from the slowly forming drop must be prevented, especially when solutions are being used. The capillary of the tip should be about 0·05 cm. in diameter, and not greater than 0·1 cm.

The following experiments represent a short study of the tensions in the system benzene-water, with and without the presence of the surface-active amyl alcohol.

* After Guye and Perrot, *Arch. Sci. phys. nat.* (Geneva), **15**, 178 (1903).

(a) Calibration of the dropping tip with water (fig. 83)

Fill a cleaned 20 ml. pipette with distilled water well above the bulb by suction at the water-pump, and fit it into a rubber bung b into which the dropping tip has already been inserted. Adjust the union so that the tip is vertical. Close the upper end of the pipette with the stopper S and wait until drops cease to form. Then carefully insert a pin between the glass rod and rubber of the stopper so that a fine air-leak is opened; adjust the valve so formed until a drop forms in not less than 20 sec. When these arrangements are satisfactory bring a small weighed flask into position immediately after a drop has fallen. To prevent evaporation loosen the clamp holding the pipette and allow the bung to rest on the rim of the collecting flask. Count 10–20 drops, and reweigh the flask.

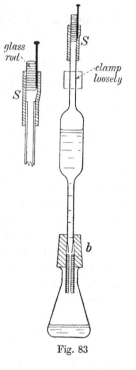

Fig. 83

(b) Add 2·5 ml. of amyl alcohol to a mixture of 50 ml. of water and 50 ml. of benzene contained in a tap-funnel.* Shake the mixture until distribution equilibrium is reached (1 min.). Allow the layers to separate fully and then run off the lower (aqueous) layer into a dry flask. Preserve the remaining benzene solution in a stoppered flask for exp. (d). Charge the dropping pipette with a sample of the aqueous layer, which contains about 1 % of amyl alcohol, and obtain the drop-weight as in (a). Retain the remainder of the aqueous solution for exp. (d).

* It is advisable to remove the grease from the tap of the funnel, by wiping with cotton-wool moistened with chloroform.

Example

(Dropping tip 6·5 mm. in diameter.)

(*a*) Distilled water: 10 drops weighed (1) 0·921, (2) 0·918 g.

Mean 0·9195 g.

(*b*) Amyl alcohol solution: 10 drops weighed (1) 0·532, (2) 0·529 g.

Mean 0·530 g.

Relative surface tension of solution $= \dfrac{0·530}{0·919} = 0·580$.

Taking $\sigma_{aq.} = 72$ dynes/cm., $\sigma_{solution} = 41·7$ dynes/cm.

(*c*) *The interfacial tension between benzene and water*

Interfacial tension is determined by estimating the drop-weight of the heavier liquid when the dropping tip is immersed in the lighter liquid (benzene in this case). It is more convenient to estimate the weight of the drop from observations on its volume than from direct weighing. The drops in many cases are much larger and take longer to form than in the surface tension experiments, owing to the Archimedean loss of weight, which must be taken into account in the calculation. The slowness of the drop formation enables the series of changes shown in fig. 82 to be well observed.

The delivery pipette of exp. (*b*) is replaced by a 10 ml. graduated tube, at the upper end of which the valve is arranged as before. The dropping tip dips below benzene contained in a small flask or beaker. After setting the dropping rate sufficiently slow, watch the water level falling in the tube, so as to be ready to read the level just as a drop falls. Read again in the same way just after 5–10 drops have fallen. Finally, take the temperature of the benzene, in order to find its density from tables.

Example

(Dropping tip as in exps. (*a*) and (*b*).)

Volume of water for 5 drops in benzene = 1·92 ml. (mean).

Volume of one drop = 0·385 ml.

Temperature   = 18·1°.

	10°	15°	20°
$D_{benzene}$	0·8838	0·8787	0·8725

D_{benzene} interpolated for $18° = 0.8760$.

Weight of 0.385 ml. of benzene $= 0.337$ g.

Weight of water drop in benzene $= 0.385 - 0.337 = 0.048$ g.

Taking the drop-weight for water in air as 0.0919 (exp. (a)),

$$\sigma_{\text{H}_2\text{O-benzene}} = 72 \times \frac{0.048}{0.0919} = 36.5 \text{ dynes/cm.}$$

(Note that $\sigma_{\text{H}_2\text{O}} - \sigma_{\text{benzene}} = 72 - 28 = 44$ dynes/cm. Cf. Antonow's rule, p. 275.)

(d) Interfacial tension of benzene-water in the presence of amyl alcohol

The dropping tube is charged with a specimen of the aqueous layer of exp. (b), and the dropping tip is immersed in the benzene layer from that experiment. The drop-weight is found as in exp. (c). The density of the benzene layer may be taken as equal to that of pure benzene.

Example

Data and calculation as in exp. (c) showed

$$\sigma_{\text{aq.-benzene layer}} = 17.4 \text{ dynes/cm.}$$

We may now draw up the following table of adhesional and cohesional work:

Interface	Cohesional work W_c ergs/cm.	Adhesional work W_a ergs/cm.
Water—water	$2 \times 72 = 144$	—
Benzene—benzene	$2 \times 28 = \overline{56}$	—
Water—benzene	—	$72 + 28 - 36.5 = 63.5$
Water—benzene (with amyl alcohol)	—	$41.7 + 28 - 17.4 = \underline{52.3}$

It is assumed, as can easily be proved by experiment, that benzene does not act as a capillary-active substance to water, nor amyl alcohol to benzene.

If the separation in the last case were really between two *polar* phases, we should expect W_a to be greater than W_a for water-pure benzene, and possibly to approach the value for water-water. The actual value is however rather less than W_c

for benzene, where hydrocarbon molecules alone are separated. This large discrepancy can only be explained by assuming, not only that the amyl alcohol in the water is con-

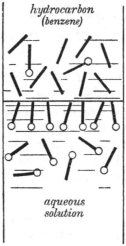

Fig. 84

centrated in the surface, but that its molecules are largely oriented with the hydrocarbon chain towards the benzene, so that the separation is effectively between a paraffin and benzene (fig. 84).

(3) Experiments to demonstrate adsorption at an oil-water interface, and the concurrent shift of chemical equilibrium*

The experiments concern the indicator equilibria (cf. p. 229):

$$A + H^+ \rightleftharpoons AH^+$$

Violet (2·0)	Green (1·0)	} Methyl-violet
Green (1·5)	Yellow (0·5)	} Malachite-green

$$A^- + H^+ \rightleftharpoons AH$$

Blue (7·4)	Yellow (6·2)	} Brom-thymol blue

---

* Deutsch, *Ber.* **60**, 1036 (1927); *Z. physikal. Chem.* **136**, 353 (1928); Freundlich, *J. Chem. Soc.* 1930, p. 164.

The figures in brackets give the pH values at which the respective colours appear fully developed to visual inspection. Only the non-ionic species A and AH can be strongly adsorbed at an oil-water interface.

(a) Add a few drops (0·2 ml.) of a solution of 0·1 g. of methyl-violet in 50 ml. of alcohol (rectified spirit) to 200 ml. of accurately $N/10$ hydrochloric acid contained in a stoppered bottle. A green solution results. Add 100 ml. of benzene, close the bottle, and shake. The contents first assume a pure blue colour, and on more vigorous shaking and further extension of the interface, a clear violet, which, however, gives place to the original green in the aqueous layer immediately the shaking is interrupted, and the dispersed oil allowed to aggregate. The interface, when viewed obliquely downwards through the colourless benzene, is seen to give a permanently violet reflexion. If, after the benzene droplets have separated, the interface is viewed upwards through the water no violet coloration is visible, a fact which emphasizes the extreme thinness of the interfacial layer.

(b) Place 200 ml. of a $2N$ solution of sodium chloride (preferably prepared from *Analar* or other pure material) in a stoppered bottle, and add a few drops (0·4 ml.) of a solution of 0·1 g. of brom-thymol blue in 50 ml. of alcohol (rectified spirit). With the aid of a glass rod moistened with the reagent, introduce just enough sodium carbonate solution to give a full and clear blue colour on shaking the bottle. Shake benzene with dilute sodium hydroxide to remove traces of acid, wash with water, separate, filter through a dry paper, and then add 100 ml. of the hydrocarbon to the bottle. On vigorous shaking the blue colour is completely replaced by the faint yellow of weakly acid solutions of the indicator. When the shaking ceases, and aggregation takes place, streamers of blue, obviously due to expulsion of (yellow) adsorbed material, appear in a striking manner descending into the aqueous layer near the rapidly contracting interface, and homogeneity is only restored to the water layer by gentle movement. A similar but less obvious effect is noticeable in exp. (a). (If the blue colour is not restored when the benzene separates, owing to remaining acidity in the benzene, a second treatment with sodium carbonate, similar to the first, will generally correct this, and lead to the expected effect; it is not necessary to separate the benzene for this second addition of alkali.) In this experiment, as in (a), the benzene remains colourless, but here the interface is permanently yellow in reflexion.

(c) On shaking 200 ml. of the yellow solution obtained by dissolving a trace of malachite-green in $0.4 N$ hydrochloric acid with 100 ml. of benzene, similar phenomena can be observed. The whole becomes of a green colour, reverting to the original yellow when at rest. The interface when viewed as described in (a) is permanently green. All the experiments (a), (b) and (c), when once set up and the bottles closed, may be repeated indefinitely. The colour of the methyl-violet tends to fade after a day or two, but the effects are restored by the addition of a further quantity of the dye solution to the oil-water system. After making such an addition the benzene layer is at first coloured violet, but one shaking suffices to extract all the dye into the acid layer, which then becomes green as before.

(4) Adsorption at the solid-liquid interface

Charcoal is the only common material (substrate) showing little discrimination in respect to the electric charge of the adsorbate; its surface will therefore adsorb both electrolytes and non-electrolytes, but the latter as a rule much more strongly than the former. The majority of substances show a definite preference for adsorption of positive or negative ions, and adsorb non-electrolytes feebly. There seems in general to be a reciprocal relation between the sign of the preferred ionic charge and the chemical character of the adsorbent —for example, SiO_2 and SnO_2 (acidic oxides) adsorb kations, while $Fe(OH)_3$ and ZnO (basic oxides) adsorb anions, but not kations; aluminium silicate (kaolin—experiment below) prefers kations, showing that in this salt the acid partner has the predominating influence. The ion adsorbed has a charge opposite to the usual charge of the adsorbent when in the colloidal state.

The less adsorbed ion will always 'antagonize' the adsorption of the other. Of all ions H^+ and OH^- seem to be the most strongly adsorbed on all surfaces, and therefore these ions least antagonize the adsorption of their partners. If a surface adsorbs kations, the adsorption will reach a maximum in the presence of OH ions, and conversely. Thus in general the amount of adsorption of electrolytes depends not only on the concentration of the solution but also on its pH.

(1) *The adsorption of (a) acetone, (b) sodium hydroxide on (i) charcoal, (ii) kaolin.*

Experiments with strongly adsorbed substances that can be made to saturate both surfaces have shown that equal weights of charcoal and kaolin have surfaces approximately in the ratio charcoal/kaolin = 10. Weigh 0·5 g. of charcoal, and 5·0 g. of kaolin (on a 'rough' balance), and place each specimen in a glass-stoppered bottle. Add to each 50 ml. of approx. $M/20$ acetone solution (4 ml. acetone per litre), and well shake until adsorption equilibrium is reached (1–2 min.). Filter both liquids through a *dry* pleated filter-paper, rejecting the first 10 ml. of filtrate (from which adsorption may have occurred on the filter-paper). As the kaolin preparation filters rather slowly it is well to cover the filter with a small clock-glass to minimize evaporation. Estimate acetone in the original solution* and in each filtrate by Messinger's method as follows.

Mix in a glass-stoppered bottle 10 ml. of the acetone solution with (about) 25 ml. of normal sodium hydroxide solution. Add with the aid of a pipette 40 ml. of $N/10$ iodine solution. Close the bottle and well shake; allow the mixture to stand 10 min. to complete the precipitation of the iodoform. Then add (about) 30 ml. of normal sulphuric acid, and titrate the liberated (excess) iodine with $N/10$ sodium thiosulphate solution.

1 ml. of $N/10$ iodine *consumed* = 0·000967 g. of acetone.

* To correct for evaporation, 50 ml. of the original solution may be submitted to a 'control' filtering, before estimation.

Example

For estimation of acetone 10 ml. samples treated with reagents as described above.

	Original solution	Filtrate	
		Kaolin	Charcoal
Titration with thiosulphate (ml.)	14·5	16·2	20·8
$N/10$ iodine consumed by acetone in 10 ml. (40 − thiosulphate titration)	25·5	23·8	19·2
Acetone in 50 ml.	123 mg.	115 mg.	93 mg.
Acetone adsorbed (by difference)		8 mg.	30 mg.

A similar set of experiments is carried out, in which 50 ml. of $N/10$ sodium hydroxide solution is used in place of the solution of acetone. Titrate with $N/20$ hydrochloric acid, using methyl orange.

Example

	Kaolin	Charcoal
NaOH adsorbed (mg.)	22	6

(The removal of NaOH by chemical double decomposition, with the production of sodium silicate, is excluded, since the silicate would be estimated as NaOH by hydrochloric acid and methyl orange.)

(2) *Adsorption displacement.*

To 300 ml. of the $M/20$ acetone solution (above) contained in a flask add 0·4 g. of benzoic acid ($M = 122$), close the flask with a cork and shake until all (or nearly all) of the acid is dissolved, thus producing a solution approx. $M/100$ to the acid and with the acetone concentration unaltered. Determine the adsorption of acetone from 50 ml. of this solution by 0·5 g. of charcoal, using the method of exp. 1. The benzoic acid does not interfere with the acetone estimation.

It will be found that the presence of the benzoic acid reduces the adsorption of acetone by about 50 %. Separate experiments with solutions of benzoic acid would show that the acid is very strongly adsorbed on charcoal, and even in $M/100$ concentration competes favourably with the $M/20$ acetone for the charcoal surface. The acid may be said to 'poison' the charcoal in respect to adsorption of acetone. Such displacement actions emphasize the fundamental character of adsorption as a *surface* condensation, since no other mode of 'sorption' by the charcoal could be so markedly influenced by the presence of very small concentrations of other substances.

(3) *The adsorption isotherm for acetone on charcoal.*

Prepare a $0.2\,M$ solution (solution 1) by mixing 15 ml. of acetone with 1000 ml. of water. Dilute 50 ml. of this solution to 200 ml. and estimate the acetone in 10 ml. of the diluted solution by Messinger's method as described in exp. 1. Calculate the concentration of the undiluted solution.

(a) Shake 100 ml. of solution 1 with 1 g. of charcoal for 2 min. Filter as in exp. 1, and estimate acetone in the filtrate, *after diluting* as above.

(b) Repeat the experiments, using successively 0.05, 0.02 and $0.005\,M$ solutions, obtained by suitable dilution of solution 1. The filtrates from these experiments will not need dilution before estimation, and 20 ml. of $N/10$ iodine solution will give sufficient excess.

Calculate by difference the acetone adsorbed (a) in each experiment. Plot $\log a$ against $\log C$, C being the concentration of acetone at equilibrium, i.e. in the filtrates. A straight line confirms the Freundlich formula

$$a = kC^{1/n}.$$

(The substitution of acetic acid for acetone in the above type of experiment cannot be recommended, as the titration of very dilute solutions of the acid even with alkali free from carbonate presents serious difficulties.)

(5) The preparation and properties of suspensoid colloids*

Suspensoid colloids, i.e. those whose micelles are composed of matter in the solid state, have been more fully investigated and are better understood than other types of colloid, mainly for the reason that they can be prepared in the laboratory from matter in the ordinary state.

All particles or micelles of suspensoid colloids have a net electric charge, due usually to the preferential adsorption of simple ions during preparation. The stability of the dispersion depends very largely on the preservation of this charge, which maintains repulsive forces between separate particles. For any given method of preparation there will be an optimum concentration of the requisite ions, corresponding to the maximum of preferential adsorption and of charge. If the concentration of electrolyte is increased beyond this optimum, e.g. by the addition of electrolyte, or decreased below it, e.g. by prolonged dialysis, preferential adsorption and the net charge diminish, and the colloid ultimately coagulates and is precipitated. The well-known rule of Hardy and Schulze states that the coagulating power of an ion (of charge opposite to that preferentially adsorbed by the micelle) increases with its valency. This is because the adsorbability, and the power to neutralize already adsorbed charge, both increase with the charge on the coagulating ions.

The rate of coagulation obviously depends on how much the electrolyte concentration differs from the optimum: if the difference is small, coagulation may be extremely slow: if large, coagulation and precipitation may appear to be instantaneous. The coagulating power of an ion thus only becomes definite when associated with a fixed coagulation period.

* A simple general account of colloid chemistry will be found in H. R. Kruyt, *Colloids*, translated by van Klooster (Chapman and Hall, 1930).

*The preparation of typical sols**

(1) *Arsenious sulphide.*

Add 1 g. of pure (*Analar*) arsenious oxide to 100 ml. of hot distilled water, contained in a flask, and keep the liquid hot on a boiling water-bath until all the oxide has dissolved ($\frac{1}{2}$–$\frac{3}{4}$ hr.). After removing the inlet tubes and bung from a flask arranged

as in fig. 85, boil in the flask 200 ml. of distilled water to expel dissolved air, and then cool to room temperature. Pour out and reserve 80 ml. of the water, and proceed to saturate the remainder with hydrogen sulphide (well washed to free from mineral acid). With the bung loose, first displace air from above the water, and then, with the bung firmly in place, well shake the water, while maintaining connexion with the gas supply until no more is absorbed. Remove the longer inlet tube with the rubber sleeve, and pour in the remaining boiled water through a drawn-out test-tube inserted through the broader inlet tube. Finally, add *slowly* in the same way the cooled solution of arsenious oxide, keeping the liquid in the flask well shaken throughout the mixing.

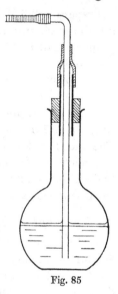

Fig. 85

The orange-yellow sol prepared in this way is slightly opalescent (owing to the Tyndall effect), but should be free from coagulum and contain only a small excess of hydrogen

---

* As_2S_3: Freundlich, *Z. physikal. Chem.* **44**, 129 (1903); Powis, *J. Chem. Soc.* (1916), p. 734. MnO_2: Bredig and Marck, *Gedenkboek aangeboden aan J. M. v. Bemmelen* (1910); *Chem. Soc. Abstr.* **2** (1911), p. 399; Ganguly and Dhar, *J. Physical Chem.* **26**, 701 (1922). $Fe(OH)_3$: Freundlich, loc. cit. supra (1903), and *Kolloid-Z.* **41**, 38 (1927); Sorum, *J. Amer. Chem. Soc.* **50**, 1264 (1928). General: *Colloid Chemistry*, Monographs, no. 16, Amer. Chem. Soc. (Chemical Catalog Co. 1928); and Freundlich, *Ber.* **61**, 2219 (1928).

sulphide, which need not be removed. Should some coagulum have formed on the surface during a too rapid final mixing the sol should be filtered through a Gooch crucible, prepared with asbestos in the usual way or through a fritted glass filter. Filtration through paper nearly always results in extensive coagulation.

(2) *Manganese dioxide.*

Prepare a $0 \cdot 6\,M$ (2 %) solution of hydrogen peroxide by diluting perhydrol (Merck) with distilled water in the volume proportion of 1 : 15. Place this solution in a burette, and add it drop by drop to 50 ml. of approx. $N/10$ potassium permanganate, keeping the mixture well shaken, until a test portion no longer leaves a pink solution on addition to a reagent solution of barium chloride (which at once precipitates the sol); about 7 ml. of the peroxide solution should be required. The micelle of the deep brown sol is negatively charged owing to adsorption of OH ions provided by the very dilute potassium hydroxide produced in the reaction between the peroxide and the permanganate. Dialysis is not necessary.

(3) *Ferric hydroxide.*

Add 5 ml. of a 10 % aqueous solution of ferric chloride to 300 ml. of briskly boiling water. The deep brown sol, with positively charged micelle, contains the hydrochloric acid produced in the hydrolysis, but this will not be found to affect materially the properties of the sol in regard to coagulation. The acid may be easily removed by dialysis.

The commercial product 'liquor ferri dialysatus', diluted ten times, may be employed in coagulation experiments, provided the original sol does not contain protective colloids.

Experiments on coagulation by electrolytes

(a) *Arsenic sulphide and manganese dioxide.*

Solutions required: $N/2$ sodium chloride, $M/10$ barium chloride and aluminium chloride.

Well mix in four labelled test-tubes respectively $N/2$ sodium chloride and distilled water as follows:

| $N/2$ NaCl (ml.) | 9 | 5 | 2·5 | 1 |
| Water (ml.) | 0 | 4 | 6·5 | 8 |

Add to each tube in turn from a dropping tube 1 ml. of the sol and immediately mix. Examine in each case for coagulation after a definite interval (1 min. or longer, as time allows).

In experiments with polyvalent ions (e.g. Ba^{++}) it is advisable to determine first the order of magnitude of the 'threshold' concentration, in the following way: Place in the first of four test-tubes 1 ml. of $M/10$ barium chloride and well mix with 9 ml. of distilled water. Take 1 ml. of the $M/100$ solution so formed and dilute with 9 ml. of water in the second tube, and so on, thus obtaining a series of concentrations in powers of 10. Add to each tube 1 ml. of the sol. More exact experiments can then be made in the region of concentration indicated.

Draw up a table of 'threshold' concentrations of the electrolytes for the period of coagulation chosen, and show that it indicates that the micelles of the two sols are negatively charged.

(b) Ferric hydroxide.

With this positively charged micelle the anionic charge is the deciding factor in coagulation. The region of the 'threshold' concentration should be explored as indicated above, using an $M/10$ solution of sodium sulphate. The effect of magnesium sulphate, barium chloride and sodium chloride should also be examined. The sol is especially sensitive to OH ion, in spite of its low charge, since this ion removes, by chemical combination, the ions (H^{+} and Fe^{+++}) adsorbed in the micelle.

(c) The co-precipitation of sols.

Sols of oppositely charged micelles usually precipitate each other on mixing, yielding a mixed coagulum. This may be tested on the above preparations.

(6) Heterogeneous or contact catalysis*

(a) Hydrogen peroxide decomposes in films adsorbed on a variety of surfaces: the metals of the platinum group are particularly active as catalysts in this way, and among metallic compounds, the polyoxides (e.g. manganese dioxide) are outstanding. The latter have the practical advantage that they are not readily poisoned, metallic catalysts being especially sensitive to traces of foreign material. Owing to their large surface, and permanent dispersion, colloidal preparations are used where possible for reactions in solution. The catalyst in the experiment described below is a sol of manganese dioxide, readily prepared *in situ* by the addition of a small quantity of potassium permanganate. The most regular results are obtained in faintly alkaline solution (borate buffer).†

(As in all experiments on reaction velocity complete cleanliness of apparatus (flasks, pipettes, etc.) is essential for success. Cleaning may be effectively carried out by leaving apparatus overnight filled with acid dichromate, or, more expeditiously, by rinsing with warm aqua regia.)

The following solutions are required:

(1) Hydrogen peroxide—about $0 \cdot 6 M$ (2 %, 6-volume). The solution should be prepared by diluting perhydrol (Merck) in the proportion of 1 vol. to 15 vol. of water. The 'inhibitors' usually added to ordinary commercial hydrogen peroxide often cause irregular results in velocity experiments.

(2) Potassium permanganate—about $N/50$.

(3) Borate buffer (Sörensen) $pH_{18°}$ $9 \cdot 24$; for formula see p. 234 (no addition of HCl or NaOH is required).

Mix in a conical flask (300 ml. capacity) 150 ml. of distilled water, 50 ml. of the borate buffer solution, and 10 ml. of peroxide solution (1). Close the flask with a plug of cotton-wool (not a cork) and place in a thermostat regulated to about 25°. When the temperature has become steady add, and well

---

* Cf. Homogeneous Catalysis, Chapter VII, p. 260.

† Bredig and Marck, *Chem. Soc. Abstr.* 2 (1911), p. 399.

mix, 3–5 ml. of $N/50$ permanganate. Take the first sample
(10 ml.) at the end of about 3 min. after addition of the catalyst, and continue to take samples at fixed intervals (5 or
10 min.) until the reaction is nearly complete. (About 1 hr.
with 3 ml., and about $\frac{1}{2}$ hr. with 5 ml., of $N/50$ potassium
permanganate.) Deliver the samples into a flask containing
dilute sulphuric acid, and titrate to a faint pink with $N/50$
potassium permanganate. (Do not lay the sampling pipette
on the bench, but keep it, when not in use, in a clean graduated
cylinder, or other suitable vessel.) To avoid the inconvenience
of the accumulation of oxygen in the sampling pipette, shake
the reaction flask just before sampling, in order to relieve the
supersaturation of the gas.

Example

Potassium permanganate to produce catalyst and for titration
$= 0 \cdot 0185 N$. Sampling (10 ml.) begun 5 min. after addition of 3 ml.
of the above permanganate. Temperature $= 25 \cdot 5°$.

Time Δt (min.)	Titration a	$\log a$	$\log \dfrac{a_0}{a_t}$	$\left(\log \dfrac{a_0}{a_t} \right) \div \Delta t$ $= k \times 0 \cdot 434$
0	23·3	1·3674	—	—
5	20·7	1·3160	0·0514	0·0103
15	16·25	1·2108	0·1566	0·0104
25	12·8	1·1072	0·2602	0·0104
35	10·2	1·0086	0·3588	0·0102
45	8·1	0·9085	0·4589	0·0102
55	6·35	0·8028	0·5646	0·0102

In another similar experiment, in which 5 ml. of $N/50$ potassium
permanganate were added to produce the catalyst, $k \times 0 \cdot 434 = 0 \cdot 0246$.

It will be seen that a good first-order constant is obtained.
This result suggests (but does not prove) that the peroxide is
only lightly adsorbed by the sol particles, and that consequently the concentration in the film is nearly proportional
to that in the bulk of the solution. A first-order reaction in the
film would thus appear also as a first-order reaction in respect
to the solution.

(b) Zero-order reactions

Aldehydes have only a slight reducing action upon dyestuffs, such as methylene blue, in homogeneous solution, but enzymes are known (Schardinger's enzyme, xanthase, etc.) which, having no action on aldehyde or dyestuff separately, induce oxidation of the aldehyde and reduction of the colouring matter (with loss of colour) when both materials are present in the same solution. Bredig showed that the enzyme may be replaced by colloidal platinum, with which catalyst formic acid may replace aldehyde. The kinetic order (zero) is not altered by these replacements.*

The following solutions will be required:

(1) Tropaeoline O (=Chrysoine=Resorcinol yellow) 1·25 g./l. = 0·005 M.

(2) Formic acid—molar (*free from chloride*).

(3) Colloidal platinum (Bredig)—1–2 mg./l. (for method of preparation, see note at the end).

Owing to the sensitiveness of colloidal metals to 'poisoning' and to coagulation by electrolytes, very great care is necessary in their handling, if the experiments are to be successful. The first requisite is completely clean apparatus. The flasks, and the 20 and 50 ml. pipettes required should be cleaned with acid dichromate. Mix in a (conical) flask 50 ml. of the tropaeoline solution with 50 ml. of the solution of formic acid (2), close the flask with a plug of cotton-wool, and place in a thermostat regulated to about 25°. Measure into a cleaned test-tube a sample (X ml., see below) of the colloidal platinum; close the tube with a plug of cotton-wool, and support it in the thermostat. Prepare in another flask (which need not be specially cleaned) a 'control' mixture of 25 ml. of the dyestuff, 25 ml. of formic acid solution (2) and $\frac{1}{2}X$ ml. of *water*. Label six test-tubes with numbers, and place in each 2 ml. of dilute hydrochloric acid (reagent concentration).

To start the reaction remove the test-tube containing the

---

* *Z. physikal. Chem.* **70,** 34 (1910): see also p. 244.

platinum from the thermostat, quickly wipe water from the outside, and then pour the colloidal solution as completely as possible into the reaction flask already containing the dyestuff and formic acid. The contents of the flask are made uniform by gentle agitation. Record the time of mixing. Remove 20 ml. samples at suitable intervals (see below) and immediately deliver the sample into the correctly numbered test-tube (care must be taken not to contaminate the tip of the pipette with hydrochloric acid). Mix the contents of the test-tubes thoroughly by shaking; the rapid coagulation of the colloid by the acid arrests the reaction. It is advisable to keep the sampling pipette in a clean graduated cylinder when not in use.

It is convenient to choose the amount of catalyst ($= X$ ml.) by preliminary experiment so that the dyestuff is completely decolorized in 15–20 min.; sampling intervals of 3 min. would then be satisfactory.

Estimation of dyestuff in the samples. The dyestuff is titrated with a solution of titanous sulphate at the boiling temperature, in the presence of excess hydrochloric acid. Owing to the inevitable volatilization of some hydrogen chloride in this process, no estimation should be undertaken until all sampling is finished. It will be found convenient to attach an L-shaped adaptor to the burette. Mix 20 ml. of the *control* solution with about 10 ml. of dilute hydrochloric acid in a conical flask of 100 ml. capacity. Boil and titrate with the Ti^{+++} solution with the liquid gently boiling throughout. The reduction of the dye is not instantaneous even at 100°. Allow the Ti^{+++} solution to flow from the burette in a slow but regular stream of drops, until the orange colour is much diminished. Shut off the burette, and wait about 30 sec., maintaining gentle boiling. Then complete the titration with slow drop-wise addition of Ti^{+++} until the liquid has only a very faint brownish yellow. Read the burette, and then add more Ti^{+++} to ascertain that the colour is permanent. For practice repeat the titration and preserve the contents of the flask as a standard tint in titrating the reaction samples. For this, pour the contents of a test-tube into the titrating flask, wash out the test-tube with about 10 ml. of dilute hydrochloric acid and add the washings to the flask. The titration of the control gives the concentration of dyestuff at zero time.

On plotting the titration values against the time a nearly straight line is obtained, indicating a zero-order reaction:

$$-\frac{dC}{dt} = \text{constant} = kC^0.$$

The simplest interpretation is to suppose that the platinum particles are surrounded, at all concentrations of the dyestuff except the very lowest, with a nearly saturated adsorption film of the dye, which is rendered active by the adsorption process. On the small remaining fraction of the surface formic acid can condense and reduce the neighbouring dye; the products of the reduction are supposed to be only feebly adsorbed, and pass into the solution. At the locality of the reaction the concentration of the dye is thus nearly constant, and the zero order of the reaction necessarily follows:

$$SO_3H \cdot C_6H_4 \cdot N{=}N \cdot C_6H_3(OH)_2 + 4H \cdot COOH$$
$$\text{tropaeoline} \qquad\qquad\qquad \text{formic acid}$$
$$\longrightarrow \underbrace{SO_3H \cdot C_6H_4 \cdot NH_2 + NH_2 \cdot C_6H_3(OH)_2}_{\text{colourless reduction products}} + 2COOH \cdot COOH$$
$$\text{oxalic acid}$$

The preparation of platinum sol by Bredig's method*

Two platinum rods, of cross-section 2–3 mm., are joined by insertion in binding screws to copper leads, over which protecting sleeves of rubber tubing are slipped as insulation in handling. 250 ml. of conductance water (for preparation, see p. 185) are placed in a glass dish (preferably of Pyrex or Jena glass) floating in a large basin of cold water. The copper leads are connected to a source of current through a low resistance (about 10 ohms for 100 V. supply) capable of taking a load of 10–12 amp. The platinum pole pieces are momentarily brought

---

* Bredig, *Z. angew. Chem.* **12**, 951 (1898); *Z. physikal. Chem.* **32**, 127 (1900).

into contact beneath the water in the dish, and then held steadily about 2 mm. apart, when an arc should strike between them, the surrounding liquid becoming rapidly darkened with the sol produced. The arc, which passes about 10 amp. when acting effectively, cannot as a rule be maintained for more than 5–10 sec. without restriking, but a total duration of at most 1–2 min. should give a sol suitable for use in exps. (b) and (c). The almost black liquid finally produced is set aside for 24 hr. and then the sol decanted from sediment; *it must not be filtered*. The platinum content can be estimated by coagulating a small specimen by the addition of dilute hydrochloric acid, evaporating somewhat, and then filtering through a weighed Gooch crucible.

(c) *An enzyme model*

The following solutions are required:

(i) Methylene-blue, 1 g. of the *pure* dyestuff per litre (solution MB).

(ii) Formic acid, approx. molar, free from chloride, as for exp. (b) above (solution F).

(iii) Colloidal platinum, 1–2 mg./l.

Prepare separately in three flasks the following mixtures, afterwards closing all the flasks with *clean* corks; for (1) and (2) use 200–300 ml. flasks, and for (3) a litre or 1500 ml. flask:

(1) 200 ml. distilled water, 20 ml. solution MB, and 50 ml. solution F.

(2) 250 ml. distilled water, 20 ml. solution MB, and 15 ml. colloidal Pt.

(3) 200 ml. distilled water, 20 ml. solution MB, 50 ml. solution F, and 15 ml. colloidal Pt.

On allowing the preparations to stand without shaking, the blue colour at once begins to fade in flask (3), but remains permanently unaltered in the other flasks, except for the loss of brightness due to the dark-coloured colloidal Pt in (2). When the colour has almost disappeared in (3), and the dye is reduced to the colourless leuco-base (equation below), add 5 drops of a 10 % solution of potassium cyanide, which destroys the activity of the catalyst. The flask is now vigorously shaken and opened from time to time to the air; the blue colour of the original dye soon returns.

$$2MB + 2H \cdot COOH \xrightarrow{Pt} \underset{\text{leuco-base}}{2(MB \cdot H)} + \underset{\text{oxalic acid}}{(COOH)_2}$$

atmospheric oxygen

It will be clear that if the atmospheric oxidation were allowed to proceed while the catalyst is active, an unlimited quantity of formic acid could be oxidized to oxalic acid by the *joint* agency of the dye and the platinum. The reaction is similar to many biological reactions in which enzymes take part, especially in the role played by the (reversible) oxidation-reduction system of the dyestuff.

(7) Experiments to illustrate special properties of certain natural colloids: the proteins

In many of their characteristic properties the proteins behave like complex amino-acids, $NH_2 \cdot CH(R) \cdot COOH$. In strongly acid (e.g. normal) solution they are present as the kationic $NH_3^+ \cdot CH(R) \cdot COOH$, and in strongly alkaline solution as the anionic $NH_2 \cdot CH(R) \cdot COO^-$. At an intermediate pH, known as the *isoelectric point* (I.E.P.), the amount of ionization of the two kinds is the same, and the net charge is zero. It appears to be generally accepted that the condition of the I.E.P. is that of a zwitter-ion $NH_3^+ \cdot CH(R) \cdot COO^-$. In this state the opposite charges must so far neutralize each other's separate effects that the protein or amino-acid behaves in respect to hydration, adsorption, etc. as though it had no charge; it may be noted that in a crystal of KCl neither ion is hydrated, although both are enclosed in water when in free solution. For most of the known proteins, as for the amino-acids, the I.E.P. lies within the limits of pH 4–6 and is com-

monly about pH 5; histidine is one of the few amino-acids
with I.E.P. at pH > 7. Owing to their amphoteric nature the
proteins can act as buffers in living systems. If a protein is
treated with a solution of pH lower than that of its I.E.P., the
surface of the micelles becomes positively charged or kationic,
and with a solution of higher pH negatively charged or anionic.
It follows that an acid-treated wool or silk will strongly adsorb
acid dyes (with a coloured anion), while alkali-treated proteins
will strongly take up basic dyes (with a coloured kation). The
dyestuffs are in fact indicators for the electrical state of the
protein surface (exp. (*a*)).

The characteristic swelling or imbibition shown by some
protein-gels when placed in water (exp. (*b*)) is a function both
of the distance of the pH from that of the I.E.P., and of the
total ionic strength of the aqueous medium; the influences
of these two factors are opposed, so that swelling reaches
maxima on each side of the I.E.P. It may be pointed out that
swelling phenomena are not confined to the amphoteric proteins, or to aqueous systems, for they are shown by the non-
amphoteric starch and agar, and by rubber in a variety of non-
aqueous solvents.

Like the swelling, the net charge and the solubility, the
viscosity of protein sols (exp. (*d*)) is minimal at the I.E.P., and
rises to maxima on each side. There are at least three distinct types of viscosity to be reckoned with in protein sols:
(*a*) electro-viscosity, due directly to the charge and the consequent electro-kinetic potential at the micelle-water interface,
(*b*) volume viscosity, due probably to changing hydration, and
hence the varying size of the micelles, (*c*) structural viscosity,
associated with changes of shape—long micelles exert more
resistance to flow than the globular or spherical form. All
these types may be influenced both by the pH and by the total
ionic strength, the two factors working as usual in opposition.

From the above short account it will be realized that the
chief properties of the proteins are regulated by the pH and the
total ionic strength of the medium in which they are placed,
and hence in order to study them properly it is often necessary

to experiment with them in buffer solutions. It is most important that the concentration of salt and therefore the ionic strength should not vary from one buffer to another.*

(a) *Amphoterism of the proteins*

Into each of four labelled test-tubes put 20 ml. of solutions of pH about 3, 4, 5 and 7 respectively. For pH $= 3 \cdot 0$ add 20 ml. of water to 50 ml. of $N/10$ acetic acid: for solutions of the higher values use the appropriate buffer mixture (tables 23–27, pp. 232–234). Place in each tube a sample of white wool (chiefly composed of keratin): the wool is more easily manipulated in the experiment which follows if it is attached to a short length of copper wire to serve as a handle. The solutions are brought to boiling temperature, and the wool thoroughly soaked and stirred in the solution, air bubbles being dislodged by rapid movement through the liquid. After 5 min. of this treatment add to each tube and well mix 10 ml. of an aqueous solution of the 'basic' dye Auramine (0·1 g./l.). Withdraw the wool after 1 min., rinse first in a spare sample of the solution with the same pH to remove surplus dye solution, and then in water. Set the dyed wool to dry on filter paper.

Repeat the preparations, substituting an 'acid' (azo-) dye, such as Xylidine-red, or a Ponceau, for the basic Auramine (the solution of the azo-dye should contain about 0·5 g./l. of water).

Using the sign + to denote a unit of colour strength in the dyed protein, the results may be tabulated as follows:

pH	'Basic' dye (= yellow *kation*)	'Acid' dye (= red *anion*)
3	0	+ + +
4	+	+ +
5	+ + +	+
7	+ + + +	?

* General references on the physical chemistry of the proteins and other lyophilic colloids: Gortner, G. F. Baker Lectures, *Selected Topics in Colloid Chemistry* (Cornell Univ. Press, 1937); Jordan Lloyd

(b) *Imbibition (swelling) as a function of* pH

1 g. of finely powdered gelatine is placed in each of six dry test-tubes. Acetate buffers of pH 5·6, 5·2, 4·6, 4·2, 4·0 with constant salt content (see table 24, p. 233) are prepared, and a solution of pH 3·0, by mixing 50 ml. of $N/10$ acetic acid with 20 ml. of $N/10$ sodium chloride. Samples of each solution are added in portions to the gelatine, which is thoroughly wetted by stirring for some minutes with a glass rod until swelling has set in uniformly. The tubes are finally filled with solution, labelled and set in a rack in order of pH and left overnight. When examined again the tops of the columns of swelled gelatine are seen to lie on a curve approximately of catenary shape, with a minimum at about pH 4·7, which is the isoelectric point of the protein.

(c) *The isoelectric 'point' (range) of casein**

Prepare a normal solution of sodium acetate by dissolving 13·6 g. of the crystalline salt to make 100 ml. of solution. Gently grind 1 g. of pure casein (the Hammarsten product is essential) in a small mortar with at first only a few *drops* of distilled water, gradually adding further *small* amounts of water and continuing the mixing until the protein has formed a creamy suspension *quite free from clots* (this condition is indispensable, and is only achieved by a gradual addition of the water—if too much water is added in one portion clotting will probably occur and will not be removed by further grinding; in such a case the preparation should be recommenced, with a fresh portion of dry casein).

Warm 25 ml. of the N sodium acetate solution, measured with a pipette, and contained in a 500–1000 ml. flask, to 50–60°; pour in the suspension, which should immediately dissolve to produce an only very faintly opalescent liquid.

and Shore, *The Chemistry of the Proteins* (Churchill, 1938); Loeb, *J. Gen. Physiol.* **3**, 85, 247, 391, 557, 667, 691 and 827 (1920–1); Kruyt, *Colloids*, chap. XIV (Chapman and Hall, 1930).

* Modified from Michaelis and Pechstein, *Biochem. Z.* **47**, 260 (1912).

Any suspension adhering to the upper parts of the flask should be mixed by a gentle swirling of the solution; even gentle shaking is to be avoided, or frothing will become troublesome. Finally, transfer with washings to a 250 ml. graduated flask and make up to this volume. The solution should not need filtering; it contains the casein dissolved in $N/10$ sodium acetate.

Label a set of test-tubes (each of capacity not less than 20 ml.) with nos. 1–8. Run from burettes the following amounts (ml.) of standard acetic acid and distilled water into the tubes respectively, and mix by shaking:

	1	2	3	4	5
$N/50$ acid	2·4	3·0	4·0	5·0	10·0
Water	15·0	15·0	14·0	13·0	8·0

	6	7	8
$N/10$ acid	3·0	12·0	15·0
Water	12·0	3·0	—

Set the tubes in order, preferably in a rack with two ranks, and arrange eight more test-tubes in the second rank opposite the first. Using a 1 ml. pipette, or a graduated dropping tube, place 1 ml. of the casein solution in each of the second set of tubes. Remove with a (fast-running) pipette 10 ml. of the acid in tube no. 1, introduce the pipette into the first tube containing casein so that the end rests on the bottom of the test-tube, before releasing the contents. In this way mixing with the casein is as rapid as possible. This rapidity of mixing is important, because the protein will pass through a range of pH conditions, which may include the isoelectric range, before the pH of the buffer produced by the acetate and acid is finally established; if during this transit the protein is given time to pass beyond the stage of opalescence to actual precipitation its condition becomes effectively irreversible. Repeat the procedure with the other pairs of tubes, each time taking 10 ml. of the acid, and keeping the tubes in correct order. The final pH of the buffers is calculated sufficiently accurately from the relation

$$pH = pK - \log\left[(\text{acid})/(\text{salt})\right].$$

(Express the acid in terms of $N/10$ solution, and remember that each tube contains 1 ml. of $N/10$ sodium acetate; take pK for acetic acid as $-\log(1\cdot8 \times 10^{-5}) = 4\cdot74$.)

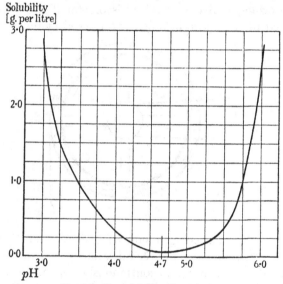

Fig. 86. The solubility of casein.*

A varying degree of turbidity is seen in the tubes immediately on mixing, and after a few minutes, close inspection will show that precipitation has taken place in some; the appearance should agree with the table below, in which amount of separation is indicated by the sign $+$:

No.	1	2	3	4	5	6	7	8
Turbidity or precipitation	0	+	++	+++	+++	+++	+	0

* Data from Sutermeister and Browne, *Casein and its Applications,* Monograph 30, Amer. Chem. Soc. pp. 76–80 (1939).

Very little, if any, difference will be observable in the appearance of the intermediate tubes 4, 5 and 6, and the range of pH in these tubes may be taken as the isoelectric range of the protein (see fig. 86).

(d) *The viscosity of protein solutions in relation to pH*

(i) *The viscosity of pure fluids (liquids or gases).* *

The definition of viscosity may be understood by reference to the principle of the Couette viscometer (fig. 87). Fluid l is placed in the space between two coaxial cylinders C_1 and

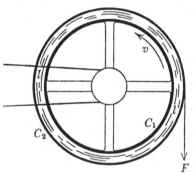

Fig. 87

C_2, the separation being d cm. If the cylinder C_1 is now given a rotation with peripheral velocity v (cm. sec^{-1}), a tangential force F must be continuously applied to the other cylinder to keep it at rest. Experiment shows that this force is directly proportional to the velocity v, and to the interfacial area A, and inversely proportional to the separation d (small compared with the diameter of the cylinders):

$$F = kAv/d.$$

When v/d and $A = 1$, $F = k = \eta$, the coefficient of viscosity. The c.g.s. unit of viscosity is termed the *poise*; it has dimensions g. cm.$^{-1}$ sec.$^{-1}$. The reciprocal $1/\eta$ is called the fluidity ϕ.

* Ward, *Trans. Faraday Soc.* **33**, 88 (1937): contribution to a general discussion on the liquid state; Andrade, *Phil. Mag.* (7), **17**, 497 and 698 (1934); *Proc. Phys. Soc.* **52**, 748 (1940); Eyring and collaborators, *J. Chem. Physics,* **4**, 283 (1936); **5**, 726 (1937); **5**, 571 (1937).

Viscous force may be considered to arise from the fact that between two fluid layers moving with different velocities there is, owing to molecular diffusion, a net transfer of momentum from the layer with the higher velocity to that with the lower. For a gas with spherical molecules this conception leads to the calculated value

$$\eta_{gas} = 0.499 \, \rho \lambda \bar{c}.$$

ρ = total mass of molecules per c.c., λ = the mean free path, and $\bar{c}$ the average velocity of the heat motion. In liquids, however, such a simple migration of molecules carrying momentum is not the sole, or even the principal, cause of viscosity, as may be seen by attempting to calculate the viscosity of molten lead from the known rate of its self-diffusion.* The theory of liquids has hardly progressed far enough for a satisfactory explanation of liquid viscosity to be given, but the trend of all modern treatment of the liquid state is to associate it with the solid, rather than with the gaseous condition. From this viewpoint viscosity appears as due to the spontaneous relaxation of shearing stresses.

In a solid a shearing stress F (fig. 88), if not excessive, is permanently balanced by internal forces, and on release of the stress the solid returns elastically to its original shape. The potential energy of the shear remains available to perform the work of restoration, and is not appreciably dissipated as heat motion. In a liquid, on the other hand, the molecular configuration can alter so as to relieve the stress, and the potential energy of the original shear is rapidly dissipated as random heat motion. Hence the force F performs continuous work on the liquid, and the layer submitted to its action slides continuously.

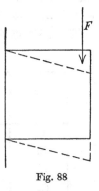

Fig. 88

---

* Andrade, *Phil. Mag.* 17, 497 and 698 (1934); Gróh and v. Hevesy, 'Self-diffusion of lead', *Ann. Physik*, 63, 85 (1920).

The rate of relief of the shear, i.e. the translatory mobility of the molecules in the liquid, will be proportional to the fluidity $\phi = 1/\eta$. It is known from experiment that the relation of viscosity to temperature is analogous to that of a reaction velocity constant to temperature (Arrhenius's equation, p. 254), except that the signs of A and B are interchanged, and viscosity *decreases* very rapidly with rise of temperature:

$$\ln \eta = B/T - A.$$

Hence it will be seen that

$$\phi = 1/\eta = \alpha c^{-B/T},$$

where $\ln \alpha = A$. It follows that the translatory movement of a liquid molecule requires an energy of activation, of average value per g.-mol. $E_1 = BR$. Various models of the liquid state have been proposed to account for, and if possible calculate, this energy of activation, but at present none is entirely satisfactory. From the data in table 30 it appears that BR (obtained from the temperature coefficient of the viscosity) is about $L_e/3$ for liquids with approximately spherical molecules, L_e being the molecular latent heat of evaporation.

Table 30. *Normal liquids*

	L_e	BR	BR/L_e	$\eta_{t.p.}$ (milli-poises)
A	1·50	0·524	0·35	2·83
N_2	1·34	0·468	0·35	3·11
CO	1·41	0·463	0·33	3·21
CH_4	2·20	0·740	0·34	2·25
O_2	1·67	0·406	0·24	8·09

(Taken from Ward's paper, *loc. cit.* supra.)

(ii) *The measurement of viscosity*

The commonest and simplest method is that of the Ostwald capillary viscometer (fig. 89), which depends upon Poiseuille's law

$$\eta = \pi \frac{hgDr^4}{8l} \frac{\tau}{V} = \text{constant} \frac{h\tau}{V}$$

for the same viscometer, and (dilute) solutions with a common
density; $h =$ head of liquid of density D; $l =$ length of capillary
of internal diameter $2r$; $V =$ the volume of liquid flowing
through the tube in τ sec.

Poiseuille's law ceases to hold when linear laminar flow gives
place to turbulence, which sets in at a certain critical velocity
dependent on the dimensions of the tube. Even when the
flow is linear within the capillary turbulence may occur at the
junction of the capillary with the wider tubes: to avoid this
the capillary should *gradually* widen. For liquids of viscosity
about one centipoise (e.g. water) the following dimensions are
safe: $2r = 0{\cdot}05$ cm., $l = 18$–20 cm.; $h = 10$–15 cm. In order to
secure the least error in h the viscometer should be supported
in a clamp in an inclined position, so that the upper bulb
is vertically above the lower (fig. 89). Owing to the extreme
sensitiveness of viscosity to temperature change an efficiently
regulated thermostat is essential. For water solutions or dispersions the viscometer is calibrated with distilled water. The
three lower bulbs shown in fig. 89 allow three separate values
of h to be used.

(For practice in operating the apparatus, the viscosities of
ethyl alcohol-water mixtures may be determined.)

(iii) *The viscosity of solutions and dispersions.**

The viscosity of a solution containing dissolved molecules
which are large compared with those of the solvent, or large
colloidal micelles is greater than that of the solvent. The
quantity $(\eta_{solution} - \eta_{solvent})/\eta_{solvent} = \eta_{sp}$, known as the *specific
viscosity*, expresses the effect on the viscosity due to the
solute. For the same type of colloidal micelle, dispersed at
low concentrations, the ratio η_{sp}/ϕ, where ϕ is the volume
fraction occupied by the micelles, varies much more slowly
than η_{sp} or ϕ. Hence if, as in the expt. below, η_{sp} decreases
solely owing to the influence of changing pH (on identical

* For detailed discussion see Alexander and Johnson, *Colloid
Science*, Clarendon Press, 1949, chap. XIII.

colloidal dispersions), then ϕ must also decrease. This result strongly suggests that the degree of solvation of the micelles is dependent on the hydrogen ion concentration in the medium.

*The determination of the viscosities of gelatine solutions.**

Warm 5 g. of granulated pure gelatine with 250 ml. of distilled water slowly (10 min.) to 45°, at which temperature the

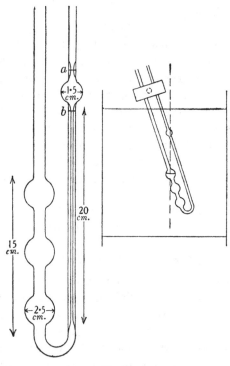

Fig. 89

mixture is kept until all the protein is dissolved (1–3 min.). During the heating the liquid should be kept in gentle motion.

* Loeb, *J. Gen. Physiol.* **3**, 99 and 827 (1920–21); Bogue, *J. Amer. Chem. Soc.* **43**, 1764 (1921); **44**, 1313 and 1343 (1922); Freundlich and Neukircher, *Kolloid-Z.* **38**, 180 (1926).

The 2 % solution thus prepared is filtered while still warm through a Gooch crucible, with asbestos mat, and is then cooled rapidly in a thermostat regulated to about 25°, where it is allowed to stand 2 hr. before use.

Clean the viscometer by drawing acetone through it, and then dry it by a current of air filtered through a plug of cotton-wool. Clamp in the thermostat with the bulbs in a vertical line (see above). With a pipette put 10 ml. of distilled water at the temperature of the thermostat into the lowest bulb; draw the water up well above the first mark a (fig. 89) with the aid of the water-pump, and then time the outflow between the marks a and b as exactly as possible. The head of water h may be decreased, and the time of outflow (τ) increased, by the addition of two further portions of 10 ml. of water to the lower bulbs. Finally, dry the viscometer as before.

Add to 50 ml. of the 2 % solution of gelatine a few drops of bromphenol-blue, when a violet-blue colour should develop. Titrate the liquid with $N/10$ HCl until the indicator just assumes a pure yellow tint, corresponding to $p\mathrm{H} = 3\cdot0$ (see table 28, p. 235); 6–7 ml. of acid should be required. Make up to 100 ml. with distilled water, thus producing a 1 % solution of gelatine. Ripen the solution by heating slowly to 45° and keeping at this temperature for 1 min. Then cool quickly by immersing in the thermostat. Introduce 10 ml. of the ripened solution into the lower bulb of the viscometer, and proceed to take the time of outflow as before when the temperature has become constant. After pouring out the bulk of the solution, remove the remainder with a current of water, and then displace the water by acetone and dry the viscometer (do not allow acetone to come into contact with any remaining gelatine solution). If time allows, the times of outflow should be found for other heads of liquid.

Repeat the procedure, including the ripening process, but substitute bromcresol-green for the first indicator, thus producing $p\mathrm{H} = 3\cdot8$ (3–4 ml. of $N/10$ acid required).

Finally, prepare isoelectric gelatine as follows—set up the half-change tint of bromcresol-green ($p\mathrm{H} = 4\cdot7$) by the method

described on p. 234. Add to 50 ml. of the 2 % solution a few
drops of $N/10$ HCl until the half-change tint is matched in the
liquid. Make up to 100 ml., ripen, and observe the outflow
time as before.*

As the viscosity of proteins is somewhat dependent on their
past history and treatment, it is wise to prove that the
ripening process has actually resulted in a stable product by
observing the outflow time of isoelectric gelatine at the beginning as well as at the end of a series of experiments.

Plot the relative viscosity $(\tau_s/\tau_{\text{water}})$ against the pH. The
following values of $\eta/\eta_0 = \tau_s/\tau_{\text{water}}$ at 24° C. are given by Loeb:

pH	η/η_0
3·0	2·90
4·0	2·10
4·7	1·35

* A more elaborate method of producing isoelectric gelatine is
given by Loeb, *J. Gen. Physiol.* **3**, 99.

Appendices

1. NOTES ON APPARATUS

(1) *Test-tubes* (ordinary and large sizes).

Suitable dimensions for the large tube ('boiling tube') recommended in many experiments are 6–7 in. by 1 in. diameter (150–175 by 25 mm.). The hard-glass type amply repays in durability and general usefulness the initial outlay.

(2) *Gas burettes*, required in the experiments of Chapters I and III (figs. 3, 13), should preferably be of the simple Ostwald pattern, capacity 250 ml. If desired, pairs of burette limbs may be purchased separately and the wooden stand be home-made.

(3) *Gas-density flasks* (fig. 3) may be readily prepared by drawing down the necks of Florence flasks (200–300 ml.) and wiring on the stout rubber connexion. To minimize the deterioration of the rubber the flasks should be stored in a dark cupboard when out of use.

(4) *Small Victor Meyer apparatus* (Cambridge pattern, fig. 7). If not home-made, the glass part, of dimensions indicated, may be inexpensively obtained to order. The bath may be made by cutting out the screw-neck from a cylindrical aluminium 'foot-warmer' or water-bottle. The cut-out portion then serves admirably as a balance-pan support for the Dumas bulb of exp. *a*, p. 19, and for other similar purposes.

(5) *Landolt pipettes*, recommended in experiments on solubility (fig. 17), may be easily blown from broken-stemmed 10 ml. volumetric pipettes.

(6) *Thermometers* for general use in freezing-point experiments should be graduated in 0·1° and range from 0 to 30°.

(7) Simple, unsilvered, straight-sided *Dewar vessels* are recommended for calorimetric experiments (Chapter v, fig. 158). The inner height may be 150 mm. and the internal diameter 30–40 mm., capacity 150–200 ml.

(8) Specifications for the construction in the laboratory of most of the *electrical apparatus* required in the experiments of Chapter VII are given in the text.

2. DATA ON THE PREPARATION, COMPOSITION, ETC. OF COMMON REAGENTS

(a) *Concentrated reagents*

Reagent	$d_{15°}$	Percentage of reagent	Anhydrous reagent in 1 ml. (g.)
H_2SO_4	1·84	98·5	1·80
HCl	1·16	31·5	0·366
HNO_3	1·42	69·8	0·991
Glacial acetic acid	1·055	100·0	1·055
NH_3	0·880	34·95	0·308

(b) *Dilute reagents*

Reagent	Preparation	$d_{15°}$	Approx. concentration	Anhydrous reagent in 1 ml. (g.)
H_2SO_4	1 vol. conc. acid to 7 vol. aq.	1·15	5 N	0·240
HCl	1 vol. conc. acid to 2·5 vol. aq.	1·05	3 N	0·104
NH_3	1 vol. 0·880 sol. to 2·5 vol. aq.	0·96	5 N	0·089
NaOH	10 g. NaOH to 90 ml. aq.	1·113	3 N	0·110
Na_2CO_3	20 g. decahydrate to 80 ml. aq.	1·079	1·5 N	0·080
NH_4Cl	20 g. salt to 80 ml. aq.	1·059	4 N	0·212
$BaCl_2$	10 g. $BaCl_2.2H_2O$ to 75 ml. aq.	1·095	N	0·109
H_2O_2	(10 vol.)	1·01	1·8 N	0·030

Index

(Crystallographic data—Chapter III—are given under this heading)

	0	1	2	3	4	5	6	7	8	9	1	2	3	4	5	6	7	8	9
10	·0000	0043	0086	0128	0170	0212	0253	0294	0334	0374	4	8	12	17	21	25	29	33	37
11	·0414	0453	0492	0531	0569	0607	0645	0682	0719	0755	4	8	11	15	19	23	26	30	34
12	·0792	0828	0864	0899	0934	0969	1004	1038	1072	1106	3	7	10	14	17	21	24	28	31
13	·1139	1173	1206	1239	1271	1303	1335	1367	1399	1430	3	6	10	13	16	19	23	26	29
14	·1461	1492	1523	1553	1584	1614	1644	1673	1703	1732	3	6	9	12	15	18	21	24	27
15	·1761	1790	1818	1847	1875	1903	1931	1959	1987	2014	3	6	8	11	14	17	20	22	25
16	·2041	2068	2095	2122	2148	2175	2201	2227	2253	2279	3	5	8	11	13	16	18	21	24
17	·2304	2330	2355	2380	2405	2430	2455	2480	2504	2529	2	5	7	10	12	15	17	20	22
18	·2553	2577	2601	2625	2648	2672	2695	2718	2742	2765	2	5	7	9	12	14	16	19	21
19	·2788	2810	2833	2856	2878	2900	2923	2945	2967	2989	2	4	7	9	11	13	16	18	20
20	·3010	3032	3054	3075	3096	3118	3139	3160	3181	3201	2	4	6	8	11	13	15	17	19
21	·3222	3243	3263	3284	3304	3324	3345	3365	3385	3404	2	4	6	8	10	12	14	16	18
22	·3424	3444	3464	3483	3502	3522	3541	3560	3579	3598	2	4	6	8	10	12	14	15	17
23	·3617	3636	3655	3674	3692	3711	3729	3747	3766	3784	2	4	6	7	9	11	13	15	17
24	·3802	3820	3838	3856	3874	3892	3909	3927	3945	3962	2	4	5	7	9	11	12	14	16
25	·3979	3997	4014	4031	4048	4065	4082	4099	4116	4133	2	3	5	7	9	10	12	14	15
26	·4150	4166	4183	4200	4216	4232	4249	4265	4281	4298	2	3	5	7	8	10	11	13	15
27	·4314	4330	4346	4362	4378	4393	4409	4425	4440	4456	2	3	5	6	8	9	11	13	14
28	·4472	4487	4502	4518	4533	4548	4564	4579	4594	4609	2	3	5	6	8	9	11	12	14
29	·4624	4639	4654	4669	4683	4698	4713	4728	4742	4757	1	3	4	6	7	9	10	12	13
30	·4771	4786	4800	4814	4829	4843	4857	4871	4886	4900	1	3	4	6	7	9	10	11	13
31	·4914	4928	4942	4955	4969	4983	4997	5011	5024	5038	1	3	4	6	7	8	10	11	12
32	·5051	5065	5079	5092	5105	5119	5132	5145	5159	5172	1	3	4	5	7	8	9	11	12
33	·5185	5198	5211	5224	5237	5250	5263	5276	5289	5302	1	3	4	5	6	8	9	10	12
34	·5315	5328	5340	5353	5366	5378	5391	5403	5416	5428	1	3	4	5	6	8	9	10	11
35	·5441	5453	5465	5478	5490	5502	5514	5527	5539	5551	1	2	4	5	6	7	9	10	11
36	·5563	5575	5587	5599	5611	5623	5635	5647	5658	5670	1	2	4	5	6	7	8	10	11
37	·5682	5694	5705	5717	5729	5740	5752	5763	5775	5786	1	2	3	5	6	7	8	9	10
38	·5798	5809	5821	5832	5843	5855	5866	5877	5888	5899	1	2	3	5	6	7	8	9	10
39	·5911	5922	5933	5944	5955	5966	5977	5988	5999	6010	1	2	3	4	5	7	8	9	10
40	·6021	6031	6042	6053	6064	6075	6085	6096	6107	6117	1	2	3	4	5	6	8	9	10
41	·6128	6138	6149	6160	6170	6180	6191	6201	6212	6222	1	2	3	4	5	6	7	8	9
42	·6232	6243	6253	6263	6274	6284	6294	6304	6314	6325	1	2	3	4	5	6	7	8	9
43	·6335	6345	6355	6365	6375	6385	6395	6405	6415	6425	1	2	3	4	5	6	7	8	9
44	·6435	6444	6454	6464	6474	6484	6493	6503	6513	6522	1	2	3	4	5	6	7	8	9
45	·6532	6542	6551	6561	6571	6580	6590	6599	6609	6618	1	2	3	4	5	6	7	8	9
46	·6628	6637	6646	6656	6665	6675	6684	6693	6702	6712	1	2	3	4	5	6	7	7	8
47	·6721	6730	6739	6749	6758	6767	6776	6785	6794	6803	1	2	3	4	5	5	6	7	8
48	·6812	6821	6830	6839	6848	6857	6866	6875	6884	6893	1	2	3	4	4	5	6	7	8
49	·6902	6911	6920	6928	6937	6946	6955	6964	6972	6981	1	2	3	4	4	5	6	7	8
50	·6990	6998	7007	7016	7024	7033	7042	7050	7059	7067	1	2	3	3	4	5	6	7	8
51	·7076	7084	7093	7101	7110	7118	7126	7135	7143	7152	1	2	3	3	4	5	6	7	8
52	·7160	7168	7177	7185	7193	7202	7210	7218	7226	7235	1	2	2	3	4	5	6	7	7
53	·7243	7251	7259	7267	7275	7284	7292	7300	7308	7316	1	2	2	3	4	5	6	6	7
54	·7324	7332	7340	7348	7356	7364	7372	7380	7388	7396	1	2	2	3	4	5	6	6	7

	0	1	2	3	4	5	6	7	8	9	1	2	3	4	5	6	7	8	9
55	·7404	7412	7419	7427	7435	7443	7451	7459	7466	7474	1	2	2	3	4	5	5	6	7
56	·7482	7490	7497	7505	7513	7520	7528	7536	7543	7551	1	2	2	3	4	5	5	6	7
57	·7559	7566	7574	7582	7589	7597	7604	7612	7619	7627	1	2	2	3	4	5	5	6	7
58	·7634	7642	7649	7657	7664	7672	7679	7686	7694	7701	1	1	2	3	4	4	5	6	7
59	·7709	7716	7723	7731	7738	7745	7752	7760	7767	7774	1	1	2	3	4	4	5	6	7
60	·7782	7789	7796	7803	7810	7818	7825	7832	7839	7846	1	1	2	3	4	4	5	6	6
61	·7853	7860	7868	7875	7882	7889	7896	7903	7910	7917	1	1	2	3	4	4	5	6	6
62	·7924	7931	7938	7945	7952	7959	7966	7973	7980	7987	1	1	2	3	3	4	5	6	6
63	·7993	8000	8007	8014	8021	8028	8035	8041	8048	8055	1	1	2	3	3	4	5	5	6
64	·8062	8069	8075	8082	8089	8096	8102	8109	8116	8122	1	1	2	3	3	4	5	5	6
65	·8129	8136	8142	8149	8156	8162	8169	8176	8182	8189	1	1	2	3	3	4	5	5	6
66	·8195	8202	8209	8215	8222	8228	8235	8241	8248	8254	1	1	2	3	3	4	5	5	6
67	·8261	8267	8274	8280	8287	8293	8299	8306	8312	8319	1	1	2	3	3	4	5	5	6
68	·8325	8331	8338	8344	8351	8357	8363	8370	8376	8382	1	1	2	3	3	4	4	5	6
69	·8388	8395	8401	8407	8414	8420	8426	8432	8439	8445	1	1	2	2	3	4	4	5	6
70	·8451	8457	8463	8470	8476	8482	8488	8494	8500	8506	1	1	2	2	3	4	4	5	6
71	·8513	8519	8525	8531	8537	8543	8549	8555	8561	8567	1	1	2	2	3	4	4	5	5
72	·8573	8579	8585	8591	8597	8603	8609	8615	8621	8627	1	1	2	2	3	4	4	5	5
73	·8633	8639	8645	8651	8657	8663	8669	8675	8681	8686	1	1	2	2	3	4	4	5	5
74	·8692	8698	8704	8710	8716	8722	8727	8733	8739	8745	1	1	2	2	3	4	4	5	5
75	·8751	8756	8762	8768	8774	8779	8785	8791	8797	8802	1	1	2	2	3	3	4	5	5
76	·8808	8814	8820	8825	8831	8837	8842	8848	8854	8859	1	1	2	2	3	3	4	5	5
77	·8865	8871	8876	8882	8887	8893	8899	8904	8910	8915	1	1	2	2	3	3	4	4	5
78	·8921	8927	8932	8938	8943	8949	8954	8960	8965	8971	1	1	2	2	3	3	4	4	5
79	·8976	8982	8987	8993	8998	9004	9009	9015	9020	9025	1	1	2	2	3	3	4	4	5
80	·9031	9036	9042	9047	9053	9058	9063	9069	9074	9079	1	1	2	2	3	3	4	4	5
81	·9085	9090	9096	9101	9106	9112	9117	9122	9128	9133	1	1	2	2	3	3	4	4	5
82	·9138	9143	9149	9154	9159	9165	9170	9175	9180	9186	1	1	2	2	3	3	4	4	5
83	·9191	9196	9201	9206	9212	9217	9222	9227	9232	9238	1	1	2	2	3	3	4	4	5
84	·9243	9248	9253	9258	9263	9269	9274	9279	9284	9289	1	1	2	2	3	3	4	4	5
85	·9294	9299	9304	9309	9315	9320	9325	9330	9335	9340	1	1	2	2	3	3	4	4	5
86	·9345	9350	9355	9360	9365	9370	9375	9380	9385	9390	1	1	1	2	3	3	4	4	5
87	·9395	9400	9405	9410	9415	9420	9425	9430	9435	9440	0	1	1	2	2	3	3	4	4
88	·9445	9450	9455	9460	9465	9469	9474	9479	9484	9489	0	1	1	2	2	3	3	4	4
89	·9494	9499	9504	9509	9513	9518	9523	9528	9533	9538	0	1	1	2	2	3	3	4	4
90	·9542	9547	9552	9557	9562	9566	9571	9576	9581	9586	0	1	1	2	2	3	3	4	4
91	·9590	9595	9600	9605	9609	9614	9619	9624	9628	9633	0	1	1	2	2	3	3	4	4
92	·9638	9643	9647	9652	9657	9661	9666	9671	9675	9680	0	1	1	2	2	3	3	4	4
93	·9685	9689	9694	9699	9703	9708	9713	9717	9722	9727	0	1	1	2	2	3	3	4	4
94	·9731	9736	9741	9745	9750	9754	9759	9763	9768	9773	0	1	1	2	2	3	3	4	4
95	·9777	9782	9786	9791	9795	9800	9805	9809	9814	9818	0	1	1	2	2	3	3	4	4
96	·9823	9827	9832	9836	9841	9845	9850	9854	9859	9863	0	1	1	2	2	3	3	4	4
97	·9868	9872	9877	9881	9886	9890	9894	9899	9903	9908	0	1	1	2	2	3	3	4	4
98	·9912	9917	9921	9926	9930	9934	9939	9943	9948	9952	0	1	1	2	2	3	3	4	4
99	·9956	9961	9965	9969	9974	9978	9983	9987	9991	9996	0	1	1	2	2	3	3	3	4

	0	1	2	3	4	5	6	7	8	9	1	2	3	4	5	6	7	8	9
·00	1000	1002	1005	1007	1009	1012	1014	1016	1019	1021	0	0	1	1	1	1	2	2	2
·01	1023	1026	1028	1030	1033	1035	1038	1040	1042	1045	0	0	1	1	1	1	2	2	2
·02	1047	1050	1052	1054	1057	1059	1062	1064	1067	1069	0	0	1	1	1	1	2	2	2
·03	1072	1074	1076	1079	1081	1084	1086	1089	1091	1094	0	0	1	1	1	1	2	2	2
·04	1096	1099	1102	1104	1107	1109	1112	1114	1117	1119	0	1	1	1	1	2	2	2	2
·05	1122	1125	1127	1130	1132	1135	1138	1140	1143	1146	0	1	1	1	1	2	2	2	2
·06	1148	1151	1153	1156	1159	1161	1164	1167	1169	1172	0	1	1	1	1	2	2	2	2
·07	1175	1178	1180	1183	1186	1189	1191	1194	1197	1199	0	1	1	1	1	2	2	2	2
·08	1202	1205	1208	1211	1213	1216	1219	1222	1225	1227	0	1	1	1	1	2	2	2	3
·09	1230	1233	1236	1239	1242	1245	1247	1250	1253	1256	0	1	1	1	1	2	2	2	3
·10	1259	1262	1265	1268	1271	1274	1276	1279	1282	1285	0	1	1	1	1	2	2	2	3
·11	1288	1291	1294	1297	1300	1303	1306	1309	1312	1315	0	1	1	1	2	2	2	2	3
·12	1318	1321	1324	1327	1330	1334	1337	1340	1343	1346	0	1	1	1	2	2	2	3	3
·13	1349	1352	1355	1358	1361	1365	1368	1371	1374	1377	0	1	1	1	2	2	2	3	3
·14	1380	1384	1387	1390	1393	1396	1400	1403	1406	1409	0	1	1	1	2	2	2	3	3
·15	1413	1416	1419	1422	1426	1429	1432	1435	1439	1442	0	1	1	1	2	2	2	3	3
·16	1445	1449	1452	1455	1459	1462	1466	1469	1472	1476	0	1	1	1	2	2	2	3	3
·17	1479	1483	1486	1489	1493	1496	1500	1503	1507	1510	0	1	1	1	2	2	2	3	3
·18	1514	1517	1521	1524	1528	1531	1535	1538	1542	1545	0	1	1	1	2	2	3	3	3
·19	1549	1552	1556	1560	1563	1567	1570	1574	1578	1581	0	1	1	1	2	2	3	3	3
·20	1585	1589	1592	1596	1600	1603	1607	1611	1614	1618	0	1	1	1	2	2	3	3	3
·21	1622	1626	1629	1633	1637	1641	1644	1648	1652	1656	0	1	1	2	2	2	3	3	3
·22	1660	1663	1667	1671	1675	1679	1683	1687	1690	1694	0	1	1	2	2	2	3	3	3
·23	1698	1702	1706	1710	1714	1718	1722	1726	1730	1734	0	1	1	2	2	2	3	3	4
·24	1738	1742	1746	1750	1754	1758	1762	1766	1770	1774	0	1	1	2	2	2	3	3	4
·25	1778	1782	1786	1791	1795	1799	1803	1807	1811	1816	0	1	1	2	2	2	3	3	4
·26	1820	1824	1828	1832	1837	1841	1845	1849	1854	1858	0	1	1	2	2	3	3	3	4
·27	1862	1866	1871	1875	1879	1884	1888	1892	1897	1901	0	1	1	2	2	3	3	3	4
·28	1905	1910	1914	1919	1923	1928	1932	1936	1941	1945	0	1	1	2	2	3	3	4	4
·29	1950	1954	1959	1963	1968	1972	1977	1982	1986	1991	0	1	1	2	2	3	3	4	4
·30	1995	2000	2004	2009	2014	2018	2023	2028	2032	2037	0	1	1	2	2	3	3	4	4
·31	2042	2046	2051	2056	2061	2065	2070	2075	2080	2084	0	1	1	2	2	3	3	4	4
·32	2089	2094	2099	2104	2109	2113	2118	2123	2128	2133	0	1	1	2	2	3	3	4	4
·33	2138	2143	2148	2153	2158	2163	2168	2173	2178	2183	0	1	1	2	2	3	3	4	4
·34	2188	2193	2198	2203	2208	2213	2218	2223	2228	2234	1	1	2	2	3	3	4	4	5
·35	2239	2244	2249	2254	2259	2265	2270	2275	2280	2286	1	1	2	2	3	3	4	4	5
·36	2291	2296	2301	2307	2312	2317	2323	2328	2333	2339	1	1	2	2	3	3	4	4	5
·37	2344	2350	2355	2360	2366	2371	2377	2382	2388	2393	1	1	2	2	3	3	4	4	5
·38	2399	2404	2410	2415	2421	2427	2432	2438	2443	2449	1	1	2	2	3	3	4	4	5
·39	2455	2460	2466	2472	2477	2483	2489	2495	2500	2506	1	1	2	2	3	3	4	5	5
·40	2512	2518	2523	2529	2535	2541	2547	2553	2559	2564	1	1	2	2	3	4	4	5	5
·41	2570	2576	2582	2588	2594	2600	2606	2612	2618	2624	1	1	2	2	3	4	4	5	5
·42	2630	2636	2642	2649	2655	2661	2667	2673	2679	2685	1	1	2	2	3	4	4	5	6
·43	2692	2698	2704	2710	2716	2723	2729	2735	2742	2748	1	1	2	3	3	4	4	5	6
·44	2754	2761	2767	2773	2780	2786	2793	2799	2805	2812	1	1	2	3	3	4	4	5	6
·45	2818	2825	2831	2838	2844	2851	2858	2864	2871	2877	1	1	2	3	3	4	5	5	6
·46	2884	2891	2897	2904	2911	2917	2924	2931	2938	2944	1	1	2	3	3	4	5	5	6
·47	2951	2958	2965	2972	2979	2985	2992	2999	3006	3013	1	1	2	3	3	4	5	5	6
·48	3020	3027	3034	3041	3048	3055	3062	3069	3076	3083	1	1	2	3	4	4	5	6	6
·49	3090	3097	3105	3112	3119	3126	3133	3141	3148	3155	1	1	2	3	4	4	5	6	6

	O	1	2	3	4	5	6	7	8	9	1	2	3	4	5	6	7	8	9
·50	3162	3170	3177	3184	3192	3199	3206	3214	3221	3228	1	1	2	3	4	4	5	6	7
·51	3236	3243	3251	3258	3266	3273	3281	3289	3296	3304	1	2	2	3	4	5	5	6	7
·52	3311	3319	3327	3334	3342	3350	3357	3365	3373	3381	1	2	2	3	4	5	5	6	7
·53	3388	3396	3404	3412	3420	3428	3436	3443	3451	3459	1	2	2	3	4	5	6	6	7
·54	3467	3475	3483	3491	3499	3508	3516	3524	3532	3540	1	2	2	3	4	5	6	6	7
·55	3548	3556	3565	3573	3581	3589	3597	3606	3614	3622	1	2	2	3	4	5	6	7	7
·56	3631	3639	3648	3656	3664	3673	3681	3690	3698	3707	1	2	3	3	4	5	6	7	8
·57	3715	3724	3733	3741	3750	3758	3767	3776	3784	3793	1	2	3	3	4	5	6	7	8
·58	3802	3811	3819	3828	3837	3846	3855	3864	3873	3882	1	2	3	4	4	5	6	7	8
·59	3890	3899	3908	3917	3926	3936	3945	3954	3963	3972	1	2	3	4	5	5	6	7	8
·60	3981	3990	3999	4009	4018	4027	4036	4046	4055	4064	1	2	3	4	5	6	6	7	8
·61	4074	4083	4093	4102	4111	4121	4130	4140	4150	4159	1	2	3	4	5	6	7	8	9
·62	4169	4178	4188	4198	4207	4217	4227	4236	4246	4256	1	2	3	4	5	6	7	8	9
·63	4266	4276	4285	4295	4305	4315	4325	4335	4345	4355	1	2	3	4	5	6	7	8	9
·64	4365	4375	4385	4395	4406	4416	4426	4436	4446	4457	1	2	3	4	5	6	7	8	9
·65	4467	4477	4487	4498	4508	4519	4529	4539	4550	4560	1	2	3	4	5	6	7	8	9
·66	4571	4581	4592	4603	4613	4624	4634	4645	4656	4667	1	2	3	4	5	6	7	9	10
·67	4677	4688	4699	4710	4721	4732	4742	4753	4764	4775	1	2	3	4	5	7	8	9	10
·68	4786	4797	4808	4819	4831	4842	4853	4864	4875	4887	1	2	3	4	6	7	8	9	10
·69	4898	4909	4920	4932	4943	4955	4966	4977	4989	5000	1	2	3	5	6	7	8	9	10
·70	5012	5023	5035	5047	5058	5070	5082	5093	5105	5117	1	2	4	5	6	7	8	9	11
·71	5129	5140	5152	5164	5176	5188	5200	5212	5224	5236	1	2	4	5	6	7	8	10	11
·72	5248	5260	5272	5284	5297	5309	5321	5333	5346	5358	1	2	4	5	6	7	9	10	11
·73	5370	5383	5395	5408	5420	5433	5445	5458	5470	5483	1	3	4	5	6	8	9	10	11
·74	5495	5508	5521	5534	5546	5559	5572	5585	5598	5610	1	3	4	5	6	8	9	10	12
·75	5623	5636	5649	5662	5675	5689	5702	5715	5728	5741	1	3	4	5	7	8	9	10	12
·76	5754	5768	5781	5794	5808	5821	5834	5848	5861	5875	1	3	4	5	7	8	9	11	12
·77	5888	5902	5916	5929	5943	5957	5970	5984	5998	6012	1	3	4	5	7	8	10	11	12
·78	6026	6039	6053	6067	6081	6095	6109	6124	6138	6152	1	3	4	6	7	8	10	11	13
·79	6166	6180	6194	6209	6223	6237	6252	6266	6281	6295	1	3	4	6	7	9	10	11	13
·80	6310	6324	6339	6353	6368	6383	6397	6412	6427	6442	1	3	4	6	7	9	10	12	13
·81	6457	6471	6486	6501	6516	6531	6546	6561	6577	6592	2	3	5	6	8	9	11	12	14
·82	6607	6622	6637	6653	6668	6683	6699	6714	6730	6745	2	3	5	6	8	9	11	12	14
·83	6761	6776	6792	6808	6823	6839	6855	6871	6887	6902	2	3	5	6	8	9	11	13	14
·84	6918	6934	6950	6966	6982	6998	7015	7031	7047	7063	2	3	5	6	8	10	11	13	15
·85	7079	7096	7112	7129	7145	7161	7178	7194	7211	7228	2	3	5	7	8	10	12	13	15
·86	7244	7261	7278	7295	7311	7328	7345	7362	7379	7396	2	3	5	7	8	10	12	13	15
·87	7413	7430	7447	7464	7482	7499	7516	7534	7551	7568	2	3	5	7	9	10	12	14	16
·88	7586	7603	7621	7638	7656	7674	7691	7709	7727	7745	2	4	5	7	9	11	12	14	16
·89	7762	7780	7798	7816	7834	7852	7870	7889	7907	7925	2	4	5	7	9	11	13	14	16
·90	7943	7962	7980	7998	8017	8035	8054	8072	8091	8110	2	4	6	7	9	11	13	15	17
·91	8128	8147	8166	8185	8204	8222	8241	8260	8279	8299	2	4	6	8	9	11	13	15	17
·92	8318	8337	8356	8375	8395	8414	8433	8453	8472	8492	2	4	6	8	10	12	14	15	17
·93	8511	8531	8551	8570	8590	8610	8630	8650	8670	8690	2	4	6	8	10	12	14	16	18
·94	8710	8730	8750	8770	8790	8810	8831	8851	8872	8892	2	4	6	8	10	12	14	16	18
·95	8913	8933	8954	8974	8995	9016	9036	9057	9078	9099	2	4	6	8	10	12	15	17	19
·96	9120	9141	9162	9183	9204	9226	9247	9268	9290	9311	2	4	6	8	11	13	15	17	19
·97	9333	9354	9376	9397	9419	9441	9462	9484	9506	9528	2	4	7	9	11	13	15	17	20
·98	9550	9572	9594	9616	9638	9661	9683	9705	9727	9750	2	4	7	9	11	13	16	18	20
·99	9772	9795	9817	9840	9863	9886	9908	9931	9954	9977	2	5	7	9	11	14	16	18	20